PENGUIN REFERENCE BOOKS

R2

A DICTIONARY OF GEOGRAPHY

W. G. Moore was educated at Burton-on-Trent Grammar School, where a film review brought him his first pen-money, and later took his B.Sc. at London University. He has spent most of his working life teaching, tutoring, lecturing, and writing, and has contributed to a variety of journals, from daily newspapers to quarterly reviews. In 1938 he published *The Geography of Capitalism*, and then wrote the geographical section of Odhams' *Complete Self-Educator*. At the start of the Second World War he was commissioned in the meteorological branch of the R.A.F., spent some years in Iraq, Persia, and the Middle East, organized the meteorological service for Persia, and completed his R.A.F. career by lecturing on Meteorology and Climatology. He has since written, in addition to this volume, *The World's Wealth* which *The Economist* described as 'a brilliantly carried out survey of the resources – animal, vegetable, and mineral – from which man gets his living', *The Soil We Live On*, and a number of series of books including *Essential Geography*, *Adventures in Geography*, and *New Visual Geography*, and contributed to the Larousse *Géographie Universelle* (English edition). He has also edited a large number of geographical filmstrips.

A DICTIONARY
OF
GEOGRAPHY

—

DEFINITIONS AND
EXPLANATIONS OF TERMS
USED IN
PHYSICAL GEOGRAPHY

—

W. G. MOORE

PENGUIN BOOKS
BALTIMORE · MARYLAND

Penguin Books Ltd, Harmondsworth, Middlesex, England
Penguin Books Inc., 3300 Clipper Mill Road, Baltimore, 11, Md, U.S.A.
Penguin Books Pty Ltd, Ringwood, Victoria, Australia

—

Published in Penguin Books 1949
Reprinted 1950
Second edition, revised and enlarged, 1952
Reprinted 1954, 1956, 1957
Reprinted with revisions 1958
Reprinted 1959, 1960, 1961
Reprinted with revisions 1962
Third edition 1963
Reprinted 1964, 1965, 1966

—

Copyright © W. G. Moore, 1949, 1962, 1963

—

Diagrams (apart from map projections)
drawn by C. R. Patrickson

—

Made and printed in Great Britain
by Hunt Barnard & Co. Ltd, Aylesbury
Set in Monotype Baskerville
Collogravure plates by Harrison & Sons Ltd

ACKNOWLEDGEMENTS

The author's thanks are due to the following for the use of material for illustrations:

To the State Highway Commission, South Dakota, U.S.A., for Plate 1.

The State Highway Commission, Oregon, U.S.A., for Plate 2.

The United States Information Service, London, for Plates 3, 6, 13, 14, 15, and 22.

The Swiss Federal Railways for Plates 4, 7, 11, and 16.

The French National Tourist Office for Plates 5 and 12.

The Vicksburg Chamber of Commerce, Mississippi, U.S.A., for Plate 8.

Trans World Airlines, for Plate 9.

The Royal Norwegian Embassy, London, for Plate 10.

Cox's Cave, Cheddar, for Plate 17.

K.P.M. Line, for Plate 18.

Canadian Pacific Railway, for Plates 19 and 21.

The Netherlands and N.I. Information Bureau, for Plate 20.

AUTHOR'S NOTE

When this book was started, it was intended to embrace the entire field of Geography, but very soon it became evident that such a volume could not be compressed within the covers of a Penguin without making some of the definitions and descriptions so brief as to be virtually useless. It was thought preferable, therefore, to limit the volume, broadly speaking, to Physical Geography, including, of course, a generous sprinkling of Climatological and Meteorological terms, and a number of unclassifiable terms which might conceivably be met during a study of Physical Geography.

For the student who consults the book purely as a dictionary, the individual definitions and descriptions should prove adequate. There may be readers, however, who are interested enough in a particular branch of the subject to seek more information than is included in a single item. For them, the headings which are suggested for further reading are indicated in the text by the use of *italics* with capital initials. A reference to ABYSSAL DEPOSITS, for example, would suggest for further reading the items on *Abyssal* and *Oozes*.

A

AA. The Hawaiian term, pronounced ah-ah, which is sometimes used for *Block Lava*.

ABLATION. The process of carrying away or removing, e.g. the melting away of the surface of a glacier, or the wearing away of the surface of a rock through the action of water.

ABORIGINES. The people who are believed to have been the original inhabitants of a region, i.e. those who were found living in the region when it was first discovered by Europeans. The term is often applied especially to Australian aborigines, who are known colloquially as 'abos'.

ABRASION. The wearing away of part of the earth's surface by the action of wind, water, or ice.

ABSOLUTE DROUGHT. In the British Isles, a period of at least fifteen consecutive days, none of which has received as much as .01 in. of rain; the definition has not been internationally accepted. See *Drought*.

ABSOLUTE HUMIDITY. The amount of water vapour present in unit volume of air, usually expressed in grams per cubic metre; the term is sometimes wrongly applied to the pressure of the water vapour in the air. See *Humidity*, *Relative Humidity*.

ABYSSAL. The term usually means 'belonging to the lowest depths of the *Oceans*'. It is assumed that in general the abyssal region has a depth greater than about 2,000 fathoms – but its limits cannot be thus defined with precision. As sunlight does not penetrate to these depths, there is no plant life. There are, however, species of animal life, whose nature varies with the depth owing to differences in temperature and in supplies of food and oxygen. Much of their food consists of organic matter which sinks from the waters above. Conditions on the ocean floor at these great depths are extremely uniform – perpetually dark, cold, and covered with a barren expanse of ooze or clay, principally *Red Clay*.

ABYSSAL DEPOSITS. The solid matter which covers the floor of the *Abyssal* region of the oceans. See *Oozes*.

ABYSSAL ROCKS. See *Plutonic Rocks*.

ACCUMULATION, MOUNTAIN OF. A mountain formed by the collection of material on the earth's surface, the best-known example being that of a volcano, or a mountain produced by the ejection of material from a volcano. Mountains of accumulation are frequently of great height and symmetrical in shape, and tend to occur as isolated peaks.

ADRET (French). A mountain slope which faces more or less equatorwards, and, being largely exposed to the sun's rays, receives considerable light and warmth during the day. The term is principally used of the Alps. The equivalent Italian term is adretto, the German Sonnenseite. See *Ubac*.

ADRETTO (Italian). See *Adret*.

ADVECTION. Transmission by horizontal movement: usually applied to the transfer of heat by horizontal movement of the air – in distinction from *Convection*, in which the movement is vertical. The most familiar example is the transfer of heat by the movement of tropical air from low to high latitudes.

ADVECTION FOG. A *Fog* which is formed when air moves horizontally over a cool surface and is thereby cooled below the *Dew Point*. It may occur over land or sea, but is most frequently caused by warm moist air from the sea moving over the land in winter after a cold spell.

AEOLIAN. Relating to, or caused by, the wind. Aeolian deposits are materials which have been transported and laid down on the earth's surface by the wind, and include *Loess*, and the sand of deserts and dunes.

AEROGRAPHY. The subject which seeks to describe the properties, dimensions, etc., of the atmosphere.

AEROLOGY. The science which treats of the atmosphere.

AFFLUENT. A *Tributary*; a stream flowing into a larger stream.

AFFORESTATION. The process of transforming an area into forest, usually when trees have not previously grown there. See *Reforestation*.

AFRICAN TORNADO. See *Tornado* (1).

AFTER-GLOW. The radiance or glow occasionally seen in the sky in mountainous regions after sunset, commencing when the sun is about 3° or 4° below the horizon.

AGGLOMERATE. A mass of coarse rock fragments or blocks of *Lava* produced by volcanic eruptions. Usually the fragments are angular, and the agglomerate differs from a *Volcanic Ash* by reason of their greater size.

AGGRADATION. A process which tends to build up the land surface by deposition of solid material in its lower areas; the term is usually applied to a river, in which case it involves the laying down of sediment on the river bed. When, along a certain stretch, the river is unable to carry all its solid loose material, for instance, the surplus is deposited at the head of the stretch, and so increases the slope of the bed. See *Degradation*.

AGONIC LINE. A line on a map joining places of zero magnetic declination, i.e. places where the magnetic compass points true north as well as magnetic north. In general, declination increases with increasing distance from the agonic line. See *Declination, Magnetic*.

AGRICULTURE. The practice of cultivating the soil in order to produce crops. In its most primitive form, as, for instance, in tropical regions like the Amazon valley and Central Africa, it simply consists in setting the plants in forest clearings; in its more advanced form, however, it demands great knowledge and skill. The type of agriculture practised in a region depends to a great extent on climatic factors, such as rainfall and temperatures, and on the kind of soil. The term is sometimes loosely used to include *Pastoral Farming* as well as the cultivation of crops.

AIR GAP. See *Wind Gap*.

AIR MASS. A mass of air more or less homogeneous in character, covering a considerable area of the earth's surface, and bounded by frontal surfaces (see *Front*). Air masses are usually classified according to the regions where they originated, e.g. tropical or polar, and according to whether they are of maritime or continental type. An air mass may thus be classified, for example, as tropical maritime, or polar continental.

AIRPORT. An aerodrome situated on one of the principal air routes.

AITOFF'S PROJECTION. An equal-area or *Homolographic Projection*, in which the whole of the earth's surface is represented on an ellipse. Land masses near to the centre of the map have a fairly accurate shape, but those near to the eastern and western margins are badly distorted; it is superior to *Mollweide's Projection*, which it resembles, however, in that the angles of intersection of the meridians and parallels are not so greatly altered towards the margins.

ALIDADE. The movable arm of a *Quadrant* or similar instrument used for reading angular distances, and carrying the indicator and the sights.

ALKALI FLAT. An alkaline, marshy area in an arid region into which one or more desert streams lead. In the dry season, when all the water has evaporated, it becomes a barren area of hard mud covered with alkali; after heavy rainfall, it becomes a shallow, muddy lake.

ALLUVIAL FAN or ALLUVIAL CONE. The deposit of sediment laid down by a swift-flowing stream as it enters a plain or an open valley, so called on account of its shape. It is in dry regions that the alluvial fan is commonest, for there the alternate drying up and flooding of the mountain streams favour its formation. The alluvial fan sometimes grows till it is many miles across, and several fans made by neighbouring streams often unite to form a continuous plain, known as a *piedmont alluvial plain*; under these conditions the depth of alluvial material may be hundreds of feet. Much of the water of a stream is absorbed by the loose material of its fan, and the soil formed by the fan is often of great agricultural value.

ALLUVIAL PLAIN. A level tract bordering a river on which *Alluvium* is deposited; it may be situated on a *Flood-Plain*, on a *Delta*, or on an *Alluvial Fan*.

ALLUVIUM. The surplus rock material consisting mainly of sand and silt, which a river has carried in suspension, and which it has been forced to deposit. Some of the most fertile land in the world consists of alluvium deposited in the deltas of the great rivers.

ALPINE. Belonging to the Alps, or, alternatively, to the higher regions of a mountain system. More strictly, the term refers to the mountainous region lying above the coniferous forests and below the permanent snow, i.e. between the *Timber Line* and the *Snowline*; the climate of this region is often known as the *Alpine climate*. In the European mountain system known as the Alps, the word 'alps' is used specifically to signify the grasslands, or Alpine pastures, which occur in the Alpine region.

ALTIMETER. A type of aneroid *Barometer*, used principally in aircraft, which is graduated to show the approximate height above the ground or above mean sea level, instead of the atmospheric pressure.

ALTITUDE. (1) Vertical distance above *Mean Sea Level*, usually measured in feet or metres.

(2) Angular distance above the horizon measured in a vertical plane, e.g. of a heavenly body.

ALTOCUMULUS. A type of medium *Cloud* which in general consists of a mass of small, relatively thin, globular patches, sometimes so close together that their edges join.

ALTOCUMULUS CASTELLATUS. A variety of *Altocumulus* cloud, in which many of the cloudlets have developed turreted tops.

ALTOSTRATUS. A type of medium *Cloud* in the form of a continuous sheet or veil, which is sometimes thin, sometimes so thick as completely to obscure the sun or moon. Normally the sun or moon shines through it indistinctly, with a gleam. It resembles thick *Cirrostratus* cloud, but does not give a halo with sun or moon. Thick altostratus cloud frequently gives a period of continuous rain.

ANABATIC WIND. A local wind caused by the flow of air during the day up valleys and mountain slopes; the slopes become heated by the sun, the air above rises in a *Convection* current, and the air of the anabatic wind moves in to take its place. An anabatic wind thus often alternates with the night-time *Katabatic Wind*. See *Valley Wind*.

ANEMOGRAM. The continuous record of wind speed and often also of direction made by an *Anemograph*; both traces show rapid fluctuations caused by eddies due to obstacles in the path of the wind at the surface.

ANEMOGRAPH. A self-recording *Anemometer* which gives a continuous trace of the speed and often also the direction of the surface wind. One of the best-known types is the Dines tube anemograph, in which the wind pressure acts upon the opening of a tube which is arranged as a vane always to face the wind; the pressure is transmitted via the tube to a float carrying a pen, and the height of the latter above the zero position of the *Anemogram* – which is fixed to a rotating drum – indicates the wind speed.

ANEMOMETER. An instrument by which the velocity and often also the direction of the wind is measured, usually in miles per hour or metres per second. The commonest type is that in which a system of cups is employed. One type, for instance, shows the rate of rotation of the cups, so that the wind velocity can be obtained, by means of a speed indicator. A very open exposure is required, in order to reduce the effects of eddies and gusts.

ANEROID BAROMETER. See *Barometer*.

ANEROIDOGRAPH. A self-recording aneroid *Barometer*. See *Barograph*.

ANNULAR ECLIPSE. See *Eclipse, Solar*.

ANTARCTIC CIRCLE. The parallel or line of latitude drawn at 66½°S. Owing to the inclination of the earth's *Axis*, the sun does not set here on one day of the year, about December 22, the southern midsummer; similarly, the sun does not rise about June 21, the southern midwinter. Within the Antarctic Circle, the number of such days increases with nearness to the Pole. At any particular time of the year, conditions are the converse of those at the *Arctic Circle*.

ANTECEDENT RIVER. A river which has cut through land that has risen in its path, and so has maintained its course: so called because it is antecedent to the present topography.

ANTHRACITE. The hard, shiny black coal which is the coal of highest rank, i.e. containing the lowest proportion of water and volatile matter and the highest fixed carbon content; it is usually considered to represent the final product in the transformation of vegetable matter through *Peat*, *Lignite*, and *Bituminous Coal*.

ANTHROPOGEOGRAPHY. The study of the distribution of human communities on the earth in relation to their geographical environment; it thus bears the same relation to Anthropology as *Biogeography* does to Biology, and as *Zoogeography* does to Zoology, etc. Some geographers assume it to be synonymous with Human Geography.

ANTICLINE. The arch or crest of a *Fold* in rock strata. See *Syncline*.

ANTICLINORIUM. A huge arch, in form resembling an *Anticline*, each limb of which consists of a number of small *Folds*.

ANTICYCLONE. A region in which the atmospheric pressure is high compared with that of adjacent areas, and which shows at least one closed *Isobar*; generally there is a series of concentric closed isobars, approximately circular or oval in shape, the highest pressure being at the centre. In the northern hemisphere the general wind circulation is clockwise round the anticyclone, in the southern hemisphere anti-clockwise. Near the centre the winds are usually light and variable, often calm, but increase in strength somewhat towards the edge of the anticyclone.

In the temporary anticyclones of temperate latitudes, quiet, settled weather conditions are characteristic, in contrast to the weather of *Depressions* or cyclones; in summer, skies are often cloudless, and temperatures relatively high, but in winter there is so much radiation that the lower layers of air become excessively cooled, and often fog also results. Such anticyclonic conditions may persist for a considerable period, for anticyclones move very slowly, and often remain almost stationary for several days.

In addition to these temporary anticyclones, there are two great belts of permanent anticyclones, situated mainly over the oceans at about 30°N. and 30°S., which move slightly northwards and southwards, and also extend and diminish, with the seasons. See *Horse Latitudes*. Seasonal anticyclones form, too, over the great land masses in winter, the most noteworthy example being the Siberian anticyclone.

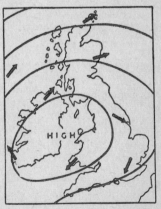

Arrangement of Isobars and surface winds in an Anticyclone (N. Hemisphere)

ANTIPODES. Two places situated on opposite sides of the earth, so that a straight line drawn through the earth from one to the other passes through the centre. London, England, and Antipodes Island, south-east of New Zealand, for example, are approximately antipodes to each other; the island is actually opposite to a point just north-west of Barfleur, on the coast of Normandy. To be exact antipodes, two places must be distant from each other by 180° of longitude, and one must be as many degrees N. latitude as the other is S. latitude. At antipodes, both the seasons and day and night are thus opposite: e.g. winter at one is simultaneous with summer at the other, and noon at one is simultaneous with midnight at the other. The term is often extended to include the whole region on the opposite side of the world from a given place.

ANTI-TRADES or COUNTER-TRADES. The winds of the upper air experienced in the regions of the *Trade Winds*, and blowing in the reverse direction to those on the surface, e.g. in the region of the NE. Trade Winds (i.e. NE. at the surface), the upper wind, the anti-trade, is SW. The anti-trades are believed to be return currents, carrying the air brought equatorwards by the Trade Winds back to higher latitudes. The term is sometimes wrongly used to denote surface winds which are opposite in direction to the Trade Winds; e.g. the SW. surface winds of the North Atlantic region, opposite in direction to the NE. Trade Winds.

APHELION. The position of the earth in its *Orbit* when it is at its greatest distance from the sun, i.e. about three million miles farther away than when it is at *Perihelion*. It reaches this position during the northern *summer*; the northern hemisphere then receives the direct rays of the sun, the days are of maximum length, and the increased *Insolation* more than counterbalances the greater distance from the sun. The term is also used for the position of any other planet or a comet when it is at its greatest distance from the sun.

APOGEE. The position in the orbit of the *Moon* or of any planet when it is at its greatest distance from the earth.

APPARENT TIME. Solar time, or time as indicated by the apparent motion of the sun, and registered, for instance, by a sundial. Apparent noon at any point on the earth's surface, then, is the instant when the sun reaches the highest point of its apparent diurnal path – or when the shadow of a vertical object is shortest.

AQUIFER or AQUAFER. A layer of rock which holds water and allows water to percolate through it.

ARABLE LAND. Land which is suitable for ploughing, as distinguished from permanent pasture and woodland. Sometimes, however, the term is applied specifically to land which is actually ploughed and on which crops are cultivated.

ARCHAEAN ERA. See *Pre-Cambrian Era*.

The group of islands, or Archipelago, of the Aegean Sea.

ARCHIPELAGO. (1) A sea studded with islands, originally applied to the sea known as the Aegean Sea.

(2) A group of islands. This is now the only meaning in general use.

ARCTIC CIRCLE. The parallel or line of latitude drawn at 66½° N. Owing to the inclination of the earth's *Axis*, the sun does not set here on one day of the year, about June 21, the northern midsummer; about December 22, the northern midwinter, the sun does not rise. Within the Arctic Circle, the number of such days increases with nearness to the Pole. At any particular time of the year, conditions are the converse of those at the *Antarctic Circle*.

ARÊTE (French). A sharp mountain ridge, often formed by the erosion of two adjoining *Cirques*. On many mountain masses, cirque erosion has taken place from several sides, leaving a series of arêtes radiating from the summit.

ARGON. An inert gas present in the atmosphere to the extent of rather less than 1 per cent of dry air.

ARID.* Deficient in *Rainfall*: usually applied to a climate or a region in which the rainfall is barely sufficient to support vegetation, sometimes – quite arbitrarily – to one in which the average annual rainfall is less than 10 inches.

ARROYO (Spanish). A stream bed, situated in a desert area, which is normally dry, but is transformed into a temporary watercourse, even a torrent, after heavy rain; when the rain has ceased, the water soon subsides, and the bed dries up again. The term is mainly used in North and South America. See *Wadi, Nullah*.

ARTESIAN WELL. A type of *Well* which normally gives a continuous flow, the water being forced upwards by hydrostatic pressure; this pressure is due to the outlet of the well being some depth below the level of the source of the water. It is often found where a basin-shaped, permeable layer of rock, such as chalk, is sandwiched between two impermeable layers, such as clay, so that rain falling on the *Outcrop* of the permeable layer will filter through and collect underground. As the water cannot escape below, the permeable layer becomes saturated to the rim of the basin, and if a well is sunk through the upper impermeable layer into it, the water flows into the well. Should the *Water Table* at the outcrop, i.e. the source of the water, be sufficiently high, the water in the well will gush out above the ground, possibly as a fountain, without the necessity of pumping. An artesian well may also be formed, for instance, on a coastal plain, where water passes downwards through a porous layer of rock, lying between two impervious layers, towards the sea. The depth of artesian wells varies considerably; some are only a few feet deep, others hundreds or even thousands of feet. Many small towns and villages rely on them for their supply of water. They are specially valuable for irrigation in semi-arid regions, such as the Great Plains of the United States and parts of Australia. The name is derived from Artois, the French province where some of the earliest artesian wells were constructed.

A type of Artesian Well: B is a permeable layer, A and C impermeable layers, W is the well.

Ås (pl. Åsar) (Swedish). In Scandinavia, an *Esker*.

ASH, VOLCANIC. See *Volcanic Ash*.

ASH CONE or CINDER CONE. The conical hill or mountain built up with the ejected material from a *Volcano*.

ASTEROID or PLANETOID. A minor *Planet*, of which there are many, revolving round the sun between the orbits of Mars and Jupiter.

The largest is about 500 miles in diameter, the smallest probably less than ten miles in diameter.

ATLAS. A collection of maps bound into a volume.

ATMOSPHERE. The envelope of air which surrounds the earth, consisting principally of a mixture of gases – mainly *Oxygen* (21 per cent), *Nitrogen* (78 per cent), *Carbon Dioxide* (.03 per cent), *Argon* (nearly 1 per cent), *Helium* and other rare gases (.01 per cent) in dry air, and a variable quantity of water vapour. The nitrogen, a chemically inactive gas, serves merely to dilute the more important oxygen. With increasing height above the earth's surface, the atmosphere becomes more and more rare, but within the layers through which adequate investigations have been carried out, the relative proportions of the gases (excluding water vapour), remain almost constant. From the viewpoint of climate and general weather conditions, the amount of water vapour present in the atmosphere, usually expressed as *Relative Humidity*, is of paramount significance; the *Temperature* of the atmosphere and the *Atmospheric Pressure* are also important.

A typical Atoll in the Pacific Ocean.

ATMOSPHERIC PRESSURE. The pressure at a point due to the weight of the column of air above that point. At the earth's surface this pressure equals about $14\frac{1}{2}$ lb per square inch, and with increasing height above the surface, the overlying column of air being shortened, the pressure decreases. The atmospheric pressure is measured by a *Barometer*, and is normally registered in *Millibars*.

ATOLL. A *Coral Reef* in the shape of a ring or horseshoe, enclosing a *Lagoon*. According to Darwin's theory, an atoll began as a fringing reef round an island, then the island became submerged owing to subsidence, leaving only the ring-shaped reef enclosing a lagoon. According to Murray's theory, an atoll was formed on the top of a plateau or hill which rose from the ocean bed to a depth at which the reef-building corals live; the outer corals of a number of colonies grew most readily and reached the surface first, thus forming an atoll enclosing a lagoon. Neither theory, however, is entirely satisfactory, for atolls are formed in both these and probably in other ways.

AUREOLE, METAMORPHIC. See *Metamorphic Aureole*.

AURORA AUSTRALIS. The light phenomenon seen in the southern hemisphere, corresponding to the *Aurora Borealis* of the northern hemisphere, being most often visible in latitudes higher than about 65° S.

AURORA BOREALIS or NORTHERN LIGHTS. The light phenomenon seen in the sky at night in the northern hemisphere, mainly in the higher latitudes. Aurora comprises an electrical discharge, and when observed as far south as in England it is almost always accompanied by a *Magnetic Storm* – but not when it is limited to higher latitudes. It assumes a variety of forms: the most stable form is the quiet arc of light, which sometimes persists with little visible change for many hours; an arc with streamers, on the other hand, is constantly changing shape.

Aurora is very rarely seen in southern Europe, is uncommon in the United States and southern England, but its frequency increases rapidly northwards; it is common, for instance, in the Orkneys and Shetlands, and a belt of maximum frequency running north of Norway and south of Greenland encircles the *North Magnetic Pole*. Its actual frequency is uncertain, for it is obscured by daylight and by cloudy skies, and, when weak, by moonlight or twilight.

AUTUMNAL EQUINOX. See *Equinox*.

AVALANCHE. A vast mass of snow and ice at high altitude which has accumulated to such an extent that its own weight causes it to slide rapidly down the mountain slope, often carrying with it thousands of tons of rock. An avalanche may thus work immense havoc, destroying villages, roads, forests in its path.

AVALANCHE CONE. The mass of material deposited where an avalanche has fallen, including snow, ice, rock, and all other objects which have been carried away by it.

AVALANCHE WIND. The high wind produced by an *Avalanche*, which sometimes causes destruction at a considerable distance from the avalanche itself.

AVEN (French). A term applied in France, mainly in the Causses region, to a *Sink Hole*.

AXIS, EARTH'S. The imaginary line, joining the North Pole and the South Pole through the centre of the earth, on which the earth rotates once in every 24 hours. It has a fixed inclination of $66\frac{1}{2}°$ to the plane of the earth's *Orbit*, and its position at any time of the year is thus parallel to its position at any other time.

AZIMUTH. The horizontal angular distance between the vertical plane passing through the observer and the Poles of the earth and the vertical plane passing through the observer and the given object; it may be measured in degrees (0°–180°) eastward or westward from the Pole, as in nautical astronomy, or in degrees (0°–360°) clockwise from true north, as in meteorology.

AZIMUTHAL PROJECTION. See *Zenithal Projection.*

B

BACKING. The anti-clockwise change of direction of a wind, e.g. from E. through NE. to N. It is the opposite change to *Veering.*

BACKWASH. The receding movement of sea water down a beach after the breaking of a wave. See *Swash.*

BAD LANDS, BADLANDS.* An elevated, arid region which is seamed and lined with deep gullies by the occasional heavy rain, normal precipitation being insufficient to support an adequate protective covering of grass or other vegetation. It is thus almost valueless for agriculture or pasture land. Unequal resistance of the rocks often leaves tall columns and platforms standing out above the surrounding land. Such a region is named after the so-called Bad Lands of the western U.S.A., where they occur particularly in western South Dakota.

BAGUIO. The *Tropical Cyclone* experienced in the Philippine Islands: a number of the tropical cyclones or *Typhoons* of the western Pacific pass over the islands.

BAHADA, BAJADA (Spanish). A *Piedmont Alluvial Plain.*

BAJADA (Spanish). See *Bahada.*

BAI-U. See *Plum Rains.*

BALLON SONDE or **SOUNDING BALLOON.** A balloon with a *Meteorograph* attached which is used for taking 'soundings' of the upper air. After being inflated with hydrogen, it is set free, and rises through the atmosphere to a height of several miles; the self-recording instruments of the meteorograph register the pressure, temperature, and humidity at various heights. Finally, the balloon bursts, the meteorograph falls to earth, being protected from injury by a light bamboo framework, and the meteorological records are obtained.

BANK. (1) A portion of the sea bed raised above its surroundings, but covered with enough water to permit navigation, e.g. the Dogger Bank in the North Sea. See *Sandbank*.

(2) The sloping ground along the edge of a river, stream, or lake.

BANKET. The conglomerate rock which forms the rich gold reef of the Witwatersrand area in the Transvaal, South Africa, so named after a type of cake made by the Boers, which it is supposed to resemble.

BANNER-CLOUD. The *Lenticular Cloud* which forms on the lee side of a high mountain summit. The enforced rise of air due to the obstruction of the mountain causes the water vapour in the air to condense and form the cloud; some distance down wind there is a descent of air, causing a re-heating and evaporation of the water drops, and there the banner-cloud disappears. Thus, although there is a constant stream of new air through the banner-cloud, the cloud itself remains stationary. Two well-known examples of a banner-cloud are the Table Cloth over Table Mountain, near Cape Town, and the cloud which forms over the Rock of Gibraltar during a *Levanter*.

GULF OF MEXICO

Sand Bars along the coast of Texas, with bays and lagoons on the landward side.

BAR. (1) The unit of atmospheric pressure, equal to one million dynes per square centimetre, and equivalent to 750.1 mm. or 29.53 in. of mercury at 0° C. in latitude 45°; for actual measurement of atmospheric pressure, the *Millibar* is in more general use.

(2) A ridge of sand and rock fragments formed across the mouth of a river or the entrance to a bay, where sediment has been deposited owing to a check being applied to the current carrying the sedi-

ment. Across the entrance to a bay, for instance, a current carrying sediment along the coast meets calmer, deeper waters which check its progress, and there it leaves its deposit; a bay may be completely sealed in this way, or an opening through the bar may be maintained by tidal movement or the movement of water from the land. Similarly, at the mouth of a river the sea causes sediment to be deposited, and estuaries are often blocked by bars. See *Nehrung*, *Spit*.

BARKHAN, BARCHAN, or BARCHANE. An isolated, crescent-shaped sand *Dune*, with the horns of the crescent projecting down wind, caused by the sand being blown round the edges as well as over the top of the heap; it is common in the desert areas of Turkestan. Sometimes several of the barkhans join together, and so lose their individual crescent shape.

BAROGRAM. The continuous record of atmospheric pressure made by a *Barograph*.

BAROGRAPH. A self-recording *Barometer*, in which a continuous trace of the atmospheric pressure is made on a *Barogram* fixed to a rotating drum actuated by clockwork. One type records the variations of pressure shown by a mercury barometer, but the type in commonest use is a modified aneroid barometer, sometimes called an *Aneroidograph*. The barograph is more useful for indicating the rise and fall of atmospheric pressure than for giving the actual pressure at any particular time.

BAROMETER. The instrument used for measuring the *Atmospheric Pressure*. In its simplest form, known as the mercury barometer, a column of mercury, normally about 30 in. or 760 mm. long, is held in a closed glass tube inverted over a bath of mercury; this column of mercury is thus balanced against the weight of the atmosphere, and its length affords a measure of that weight, i.e. of the atmospheric pressure. When the pressure is increasing, the level of the column rises, when decreasing, the level falls; a length of column much greater than 760 mm. is regarded as representing 'high' pressure, while much less than 760 mm. represents 'low' pressure. Atmospheric pressure is now usually measured in *Millibars*, however, 1013 mb. being equivalent to 760 mm. of mercury.

Another type is the aneroid barometer, which consists of a hollow metallic box or capsule nearly exhausted of air, with its lid fluted so as to render it flexible; a lever is mounted with a needle to indicate on a graduated dial the variation in pressure on the lid. The aneroid barometer is used where convenience of handling is of greater importance than accuracy, as in the domestic wall or table barometer, and in the *Barograph*.

BAROMETRIC PRESSURE. See *Atmospheric Pressure*.

BAROMETRIC TENDENCY. See *Pressure Tendency*.

BARRAGE. A large structure erected across a river in order to store water, usually for irrigation. If hydro-electric power is also produced, the structure is generally termed a *Dam*.

BARRIER REEF. See *Coral Reef*.

BARROW or TUMULUS. A mound, dating from prehistoric times, built over a burial ground: commonly found in western Europe, including the British Isles.

BARYSPHERE, BATHYSPHERE, or CENTROSPHERE. The inner portion of the earth below the *Lithosphere* or outer crust. Conflicting opinions are held as to its exact nature, but the materials of which it is composed are known to be dense, and pressure and probably temperature in the barysphere are high; according to tidal observations, at any rate, the earth is as rigid as a ball of steel of the same size.

BASALT. A dark-coloured *Igneous Rock*, of which there is a large number of types, formed by the solidification of *Lava*. Basalts are distributed all over the world, and sometimes occur in huge masses.

BASE-LEVEL. The lowest level to which a stream can wear its bed. The length of time taken for the bed to reach base-level depends on the rate of erosion; generally speaking, this will be less for a large stream than for a small one, and will be less over weak rock than over resistant rock. The permanent base-level is the level of the sea; a lake provides a temporary base-level, but the sediment deposited in it by the stream destroys its effect.

BASIN. (1) A region in which the strata or layers of rock dip in all directions towards a central point; an inverted *Dome*.

(2) A *river basin* is the total area drained by the river and its tributaries.

(3) A *lake basin* is the basin filled by the water of the lake.

(4) Any hollow or trough in the earth's crust, whether filled by water or not.

BATFURAN. The season of 'sea open' in the Arabian Sea: the winter, or period of the NE. *Monsoon*, when the dominant feature of the pressure system over Asia is the Siberian anticyclone, winds being comparatively light over the Arabian Sea, and therefore favourable to native shipping.

BAT HIDDAN. The season of 'sea closed' in the Arabian Sea: the summer, or period of the SW. *Monsoon*, when off-shore winds from the Somaliland region cause the Arabian Sea to be so stormy that native shipping is laid up.

BATHOLITH or BATHYLITH. A dome-shaped mass of rock formed by the intrusion of *Magma*, often consisting of *Granite*. It is usually much larger than a *Laccolith*, often extending over hundreds of square miles, and appears to have *replaced* rather than *displaced* the invaded rocks. Batholiths are revealed when the overlying rocks have been worn away. They form the sub-structure to mountain ranges, and appear to continue downwards to enormous depths.

BATHYAL ZONE. See *Continental Slope*.

BATHYSPHERE. See *Barysphere*.

BAY. A wide indentation into the land formed by the sea or by a lake.

BAYOU. A marshy creek or offshoot to a river or lake; occurring in flat country, it remains swampy owing to floods, river seepage, and lack of drainage. The term is chiefly used in the southern United States, where it often signifies an *Ox-Bow Lake*.

BEACH. The strip of land or terrace bordering the sea, usually recognized as that part which lies between high and low water marks, and formed by the action of the sea; on exposed coasts it consists of boulders or pebbles, on protected coasts of sand. A cliff standing on the landward side of the beach is gradually worn down by the sea, and the beach is thus widened, but when the waves can no longer reach the cliff, erosion by the sea practically ceases, and the beach no longer widens. See *Raised Beach*.

BEARING. The horizontal angle between the direction of an object and the *Meridian* through the observer, measured in degrees (0°–360°) clockwise from true north; this is the *true bearing* of the object. The *magnetic bearing* is the corresponding angle measured clockwise from the magnetic north. The bearing of an object may be given in terms of the compass points, e.g. N. 40° W., or 40° W. of N., or approximately as NW., etc.

BEAUFORT SCALE. The series of numbers devised by Admiral Beaufort at the beginning of the nineteenth century to differentiate approximately between various wind strengths: e.g. 0 on the Beaufort Scale represents a calm, the air being practically motionless, while at the upper end of the scale 12 represents a hurricane, the surface wind speed being greater than seventy-five miles per hour. Between these two extremes, the numbers indicate the intervening wind speeds, which can usually be estimated approximately by such natural phenomena as smoke drift, leaves and twigs in constant motion, small waves on inland water, violent movement of trees, slight structural damage to buildings. The various wind velocities to which the numbers on the Beaufort Scale are equivalent have been internationally agreed upon.

BEDDING PLANE. The surface which separates one layer of *Sedimentary Rock* from another.

BEDROCK. The solid rock beneath the loose material, or soil, and subsoil, with which most of the land surface of the earth is covered. It is sometimes several hundred feet beneath the surface, but is usually found at a much smaller depth; in places, especially on steep slopes, it has no soil cover at all.

BEHEADING. See *River Capture*.

BENCH MARK. In surveying, a mark, usually cut in stone, taking the form of an arrow with a horizontal bar across its apex, and used as a reference mark in the determination of altitudes. On a British Ordnance Survey map it is indicated by the letters B.M. and the height in feet.

BENTHOS. The sedentary, bottom-dwelling marine organisms of plant and animal origin, as distinct from the *Plankton* and the *Nekton*; they include seaweeds, molluscs, and corals.

BERGSCHRUND (German). The gap or *Crevasse* left round the upper rim of a glacier or a snowfield, as the ice or snow moves downwards; it usually provides a difficult obstacle to any attempt to climb from the snowfield to the mountain peaks beyond. The French term for bergschrund is rimaye.

BERG WIND. The warm, dry wind of *Föhn* type experienced in South Africa, especially on the generally cool western coast, being most frequent in winter, when the plateau of the interior is covered by a strong anticyclone and pressure is low over the ocean. The winds thus blow outwards from the plateau, affecting all margins, and become heated by descent. The temperatures may rise above 100° F., and thus temporarily exceed those of midsummer. The berg wind may continue for two or three days, giving oppressive weather, and sometimes causing severe damage to crops; its direction and the season of its occurrence vary with situation.

BIGHT. An indentation in the sea coast similar to a *Bay* but either larger or with a gentler curvature: e.g. the Great Australian Bight.

BILL. A narrow headland, or small *Peninsula*.

BIOCLIMATOLOGY. The study of climate in relation to life and health, one of its objects being to determine the climatic conditions most favourable to human habitation, especially for invalids, and to define the areas where such climates exist.

BIOGEOGRAPHY. The study of the geographical distribution of plants and animals over the globe, being usually limited to the land surface; it is divided into *Phytogeography* and *Zoogeography*.

BIOSPHERE. A term sometimes applied to that portion of the *Earth* occupied by the various forms of life, being additional, for purposes of classification, to the three main physical zones, the *Lithosphere*, the *Hydrosphere*, and the *Atmosphere*.

BISE (French). A cold, dry wind from NW., N., or NE., experienced in southern France and Switzerland, often accompanied by heavy clouds.

BITTER LAKE. A lake whose waters contain large quantities of chemical salts, such as carbonates of sodium, calcium, magnesium, in solution. See *Salt Lake*.

BITUMINOUS COAL. The shiny black coal familiarly known as household coal, which according to classification of coals is of higher rank, i.e. has a higher fixed carbon content, than *Lignite*, but lower than *Anthracite*.

BLACK COTTON-EARTH. The fertile volcanic soil of NW. Deccan, in India, where cotton is largely cultivated; the soil is very retentive of moisture, which it holds long after the rains have ceased.

BLACK EARTH. A fine, fertile soil, black or dark brown in colour, covering an extensive area of the southern U.S.S.R., north of the Black Sea, and part of Hungary and Rumania, consisting of *Loess* mixed with a large proportion of *Humus* and some lime. The black earths of the U.S.S.R., largely in the Ukraine, form much of the country's richest agricultural land. The soil is so rich in plant foods that it will take crops for long periods without the addition of fertilizers, and has no superior for production of cereals, etc. It is often known as chernozem (Russian). A similar soil also covers a belt of land extending from Saskatchewan, Canada, through N. Dakota to Texas, in the U.S.A. It is associated with a natural cover of grass.

BLIND VALLEY. In a *Karst Region*, a valley which ends suddenly where its stream disappears underground. The term is sometimes understood to signify a *Polje* caused by erosion.

BLIZZARD. A storm of powdery snow, sometimes with small ice crystals driven along by an abnormally high wind, reducing visibility practically to zero. Some of the snow falls from the clouds, but much of it is swept up from the ground. In Canada and the northern U.S.A. the blizzard, a special type of *Cold Wave*, often disturbs the calm of the winter anticyclone, coming with a northerly wind of gale strength in rear of an eastward-moving depression, and is most dangerous over the open prairies. In polar regions blizzards are frequent in some localities, at times excessively severe, and may continue for some days. Adelie Land, in Antarctica, for instance, is specially subject to them, and has been called 'the home of the blizzard'. A sudden rise of temperature occurs at the commencement of a blizzard in winter, as the surface *Inversion* of temperature is destroyed; in summer the temperature falls with a blizzard.

BLOCK LAVA. A sheet of *Lava* which has solidified so that its surface is a mass of rough, jagged blocks; the Hawaiian term *Aa* is sometimes used.

BLOCK MOUNTAIN. A mountain mass formed by the uplift of land between *Faults* or by the subsidence of land outside the faults. See *Horst*.

BLOOD-RAIN. Rain which is tinted a reddish colour, leaving a red stain on the ground. The colour is due to the imprisonment, in the drops, of dust particles which have been carried along in the upper air from a desert, often for long distances. It has been most often observed in Italy, the dust having originated from the Sahara, but has occurred as far afield as Great Britain.

BLOSSOM SHOWERS. The rains of March to May in the coffee-growing districts of the monsoon region of SE. Asia; also called 'mango showers'.

BLOWHOLE. A hole near to the sea-shore which has been formed in the roof of a cave, and through which air and possibly water are forced by the rising tide.

BLUE MUD. A marine deposit containing much decaying organic matter and also finely-divided iron sulphide; the latter gives it its characteristic dark blue or slaty colour. It covers much of the sea bed of the Pacific Ocean, the Arabian Sea, and the Bay of Bengal. See *Green Mud, Red Mud.*

BLUFF. A headland or cliff with a bold and almost perpendicular front, usually applied to the steep slopes bordering a river. These bluffs are often formed by the action of the river in cutting into the valley sides.

BOCAGE (French). A type of farming country which is divided by hedges and trees into small fields, applied especially to NW. France.

BOG. An area of soft, wet, spongy ground, consisting chiefly of decayed or decaying moss and other vegetable matter. It oftens forms in shallow, stagnant lakes or ponds, and is largely produced by sphagnum moss; the latter spreads out from the shores, floating on the surface, and gives a deposit of vegetable matter on the bottom. In time the sphagnum moss will cover the entire surface of a lake or pond, a thick mass of decaying vegetation lying below; this is often called a *quaking bog,* for it quakes under foot. Eventually, peat is evolved. Bogs are also formed on cold, damp mountain surfaces. See *Swamp.*

BOGAZ. In a *Karst Region,* a long narrow chasm, enlarged by solution of the limestone, into which a surface stream empties.

BOHOROK. The *Föhn* type of wind experienced on the plains of Deli, Sumatra, during the NE. monsoon; being warmed and dried by descent on the lee side of the mountains, it may cause much damage to such crops as tobacco.

BOLSON. A basin of interior drainage in an arid or semi-arid region, whose floor is tending to be filled by a number of *Alluvial Fans* around its flanks. The term is chiefly used of Mexico and the SW. United States.

BONNE'S PROJECTION. A *Map Projection* which is a modification of the simple (one-standard) *Conical Projection,* and resembles the latter in appearance except that the meridians are curved. As in the simple conical projection, the central meridian is straight and is divided truly, the selected standard parallel is divided truly, and the other parallels at their true distances apart are drawn as concentric circles. The modification consists in the fact that *all* parallels are divided truly, i.e. the distances between the meridians along all parallels are made equal to those distances on the globe, and the meridians are formed by drawing curves through the corresponding points on each parallel – not, as in the simple conical projection, by joining the vertex to the points of division of the standard parallel. Thus the projection is *Homolographic* or equal-area, and is on this account popular; it is not, however, suitable for the polar regions, and the more limited its extension in longitude the better.

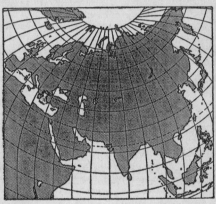

Bonne's Projection.

BORA. The cold and often very dry northerly or north-easterly wind experienced along the eastern coast of the Adriatic Sea and in northern Italy, mainly in winter; it occurs when atmospheric pressure is high over central Europe and the Balkans, and low over the Mediterranean. As it thus blows from the former area, it usually gives clear skies and cold, dry weather, but if it is associated with a depression over the Adriatic, it may be accompanied by heavy cloud and rain or snow. If the continental anticyclone is well established, the bora may continue for several days. It often blows with great strength, with violent gusts and squalls, sometimes exceeding 100 miles per hour. It corresponds to the *Mistral* of southern France.

BORE. A high tidal wave experienced in a narrow river estuary, advancing upstream like a wall of water many feet high. It is produced by the sudden retardation of the normal tidal wave, when it reaches shallow water and meets the river current; this causes the water to pile up across the river, the crest falls over, and the water moves along like a large broken wave. A bore is experienced in the River Severn, England, where at the *Spring Tides* it is often three or four feet high, and also in such rivers as the Seine, Hooghly, and Yangtse-kiang. It is sometimes known as an eagre or eger.

BOSS. A term sometimes applied to a *Batholith* of relatively small size, especially one which is approximately circular in outline.

BOTTOM. In the U.S.A., an *Alluvial Plain*.

BOULDER CLAY. The mass of rocks and finely ground *Rock Flour* dragged along in the lower part of the ice of a glacier, and left behind when the ice melts. It is usually a tough, unstratified clay, loaded with stones, and it may be several feet thick, though only the upper few inches have been modified to form a soil. Some so-called boulder clay, on the other hand, contains little or no clay, but perhaps sand, and no boulders. Its composition, which is very variable, depends on the rocks from which it came. Boulder clay is well distributed over northern Europe, including the British Isles (except the extreme south), being the remains of the ground *Moraines* of glaciers of the *Ice Age*. Boulder clay is also sometimes known as till.

BOURNE. An intermittent stream, applied especially to the chalk regions of southern England. The bed is usually dry except in winter, when the *Water Table* rises above its level.

BRAE. In Scotland, a hillside or bank overlooking a valley.

BRAIDED STREAM. A network of small shallow interlaced streams derived from a single stream. It was formed after the original stream had deposited sediment, dividing and forming new channels, and then these channels had joined up, divided, and joined up many times.

BRAVE WEST WINDS. The westerly *Planetary Winds* of the temperate region of the southern hemisphere, which, owing to the comparatively slight obstruction by land masses, blow with great force and regularity. Striking the western mountainous coasts of southern Chile, Tasmania, and New Zealand, they bring heavy rainfall, amounting in many areas to over 100 inches, while on the eastern side the rainfall is often less than 20 inches. See *Roaring Forties*.

BREAKER. A wave breaking into foam as it advances towards the shore, where the water becomes so shallow that there is insufficient to complete the wave form; or a wave breaking against a rock or other solid object.

BREAKWATER. A barrier built into the sea in order to break the force of the waves, and thus to serve as a protection against them.

BRECCIA (Italian). Rock consisting of angular fragments cemented together in a matrix; it may consist of any type of material, and is formed in a large variety of ways.

BREEZE. The term is usually applied to a current of air which is too light to be called a wind. See *Land Breeze*, *Sea Breeze*.

BRICKEARTH. A finely-grained deposit which overlies the gravel in various parts of the London basin, England, and is so called because of its use in brick-making.

BRICKFIELDER. The hot wind experienced in south-eastern Australia caused by the movement of tropical air southwards, especially during the summer, and often bringing clouds of dust. Hot spells many days in duration, with temperatures daily exceeding 100° F., often occur while the brickfielder blows. The brickfielder precedes the *Southerly Burster*.

BRIDGE, NATURAL. See *Natural Bridge*.

BRITISH SUMMER TIME or B.S.T. The standard of time, first introduced in 1916, in use in the British Isles for a period during the summer, being one hour in advance of *Greenwich Mean Time* or G.M.T.; 7 o'clock B.S.T. thus corresponds to 6 o'clock G.M.T.

BROAD. A wide stretch of water, either forming part of the course of a sluggish river, or bordering the river near to its estuary, particularly in East Englia, England.

BROCKEN, SPECTRE OF THE. The greatly magnified shadow of an observer thrown on to a bank of cloud or a mist in a mountainous region: so called after the Brocken, a peak in the Harz Mountains, Germany.

BRONZE AGE. The period when men used implements and weapons made of bronze, an alloy of copper and tin. It does not necessarily denote a fixed chronological period in history, but marks a stage of human culture through which many peoples of the world passed, in their progress from Stone Age to Iron Age.

BROOK. A small stream or rivulet.

BROWN COAL. See *Lignite*.

BRÜCKNER CYCLE. The cycle of various climatic and other natural phenomena, recurring very irregularly, which include periods of warm and dry years alternating with cold and damp years, the advance and retreat of Alpine glaciers, the level of the Caspian Sea and its inflowing rivers, and a number of meteorological data. This cycle had been known before Brückner, but he rediscovered it in 1890, and calculated its average length at approximately thirty-five years; individual cycles, however, range in length from twenty-five to fifty years.

BUFFER STATE. A state situated between two or more powerful states; usually independent, and serving the purpose of helping to prevent war between them.

BUND. Chiefly in India, an artificial embankment, dyke, or dam. In certain Chinese ports which have or have had commercial links with England, e.g. Shanghai, the Bund is the name given to the water-front.

BURAN. The cold, fierce, northerly or north-easterly wind experienced in Siberia and Central Asia, mainly during the winter in rear of a depression, breaking the comparative calm of anticyclonic conditions, carrying with it snow and ice particles; it thus corresponds to the Canadian *Blizzard*. It reaches gale force, often blows at temperatures of 20° or more below zero F., and is thus dangerous to both human and animal life, especially on the open steppes, as the blizzard is on the prairies. A strong north-easterly wind in summer is also called a buran. The winter buran is also termed the *Purga*.

BURN. In Scotland, a brook or small stream.

BUSH. A term used mainly in Australia and South Africa to signify a region covered with forest or scrub not yet cleared for cultivation; or, more loosely, any wild and uncultivated region, even though not wooded.

BUTTE. A flat-topped hill, produced when hard strata of rock overlie weaker layers, and protect them from being worn down. The butte is similar to but smaller than a *Mesa*, and is often produced from a mesa when the latter is reduced in size through dissection and erosion: it is characteristic of the arid plateau region of the western United States.

BUYS BALLOT'S LAW. The law, enunciated by Buys Ballot in 1857, which states that if an observer in the northern hemisphere stands with his back to the wind, the atmospheric pressure will be lower to his left hand than to his right, the reverse being true for the southern hemisphere. Expressed in a different way, the law states that in the northern hemisphere the winds move anti-clockwise round centres of low pressure and clockwise round centres of high pressures, the reverse being true, again, for the southern hemisphere.

C

CACIMBO. The heavy mists off the coast of Angola, occurring in the mornings and evenings, from which drizzle sometimes falls. They are characteristic of the dry season, when they cause the air to become extremely moist, and die out in the rainy season. They correspond to the *Smokes* of the Guinea Coast.

CAINOZOIC or KAINOZOIC or CENOZOIC ERA. The New Life Era, the fourth of the five major sub-divisions of the geological scale of time: that period in the earth's history extending from the close of the *Mesozoic Era* to the beginning of the *Quaternary Era*. Some geologists, however, omitting the Quaternary Era from the geological scale of time, consider the Cainozoic Era to extend to the present. The Cainozoic Era is the era when mammals were widely distributed over the earth, and the abundant vegetation was beginning to resemble that of today. It is sometimes known as the Tertiary Era.

CAIRN. A heap of stones, in ancient times erected for memorial purposes, but more recently often erected as a landmark.

CALDERA.* A large basin-shaped *Crater* bounded by steep cliffs, often formed by the subsidence of the top of a volcanic mountain, and sometimes occupied by a lake. If the volcano is not yet extinct, one or more active cones or craters may exist within the caldera. In the largest known caldera (fourteen miles by ten miles), that of Aso, in Japan, for example, there is an active volcano. The best-known caldera in North America is Crater Lake, Oregon, in which stands a small island which is a volcanic cone formed after the original volcanic summit had sunk.

CALENDAR. A system by which time is divided into fixed periods, known as years, months, days, etc. The year, the month, and the solar day are natural periods of time which have always been used in the preparation of a calendar; they correspond respectively with the period of the earth's revolution round the sun – 365 days, 5 hours, 48 minutes, 46 seconds, the period of revolution of the moon round the earth, and the period of rotation of the earth on its axis. One obstacle to the production of a calendar is the fact that the year is not an exact multiple of the time of the moon's revolution round the earth or of the solar day.

To give a solar year of 365¼ days, three consecutive years of 365 days and a fourth or leap year of 366 days are arranged. But such a year is 11 minutes 14 seconds longer than the true year. The Gregorian Calendar (New Style) now in use corrects this error by arranging that the centurial years shall be taken as leap years only when they are exactly divisible by 400. Thus 1700, 1800, and 1900 were ordinary years, but 2000 will be a leap year. The error is reduced by this means to less than one day in 3000 years.

CALINA. The haze which occurs during July and August in Mediterranean lands, reduces visibility, and turns the blue sky into a dull grey colour; it is caused by dust particles which have been swept up by strong winds.

CALMS OF CANCER. The region of calms and light winds within the belt of high pressure near the *Tropic of Cancer*. See *Horse Latitudes*.

CALMS OF CAPRICORN. The region of calms and light winds within the belt of high pressure near the *Tropic of Capricorn*. See *Horse Latitudes*.

CALVING. The detachment of an *Iceberg* from the front of a glacier, when the latter reaches the sea, or the detachment of a portion of an iceberg when the latter is afloat.

CAMPOS. The tropical grasslands or *Savanna* of Brazil, situated south of the equatorial forests of the Amazon basin.

CANAL. An artificial watercourse, usually cut in order to facilitate the cheap transport of bulky goods by barge, when the time taken in transit is not an important factor. A canal is often cut between two important rivers, in order to improve the waterway systems of the area, e.g. the Rhine-Marne Canal. Canals are also often constructed to facilitate irrigation in regions which are deficient in rainfall. See also *Ship Canal*.

CANCER, TROPIC OF. See *Tropic of Cancer*.

CANYON or CAÑON.* A *Gorge*, relatively narrow but of considerable size, bounded by steep slopes. It has often been formed by a river cutting through the soft rocks of an arid region; the scantness of the rainfall prevents denudation of the canyon walls, and so maintains their steepness. The walls of a large canyon, however, rarely approach the vertical, and their irregularity of slope is due to inequalities in the hardness of the rock. For instance, the Grand Canyon of the Colorado River in the U.S.A., the largest and best-known canyon in the world, has a depth of over a mile, in places, but although it is very narrow at the bottom in these places, it is often eight to ten miles wide at the top. A smaller ravine through which a tributary flows into the main river is known as a *side-canyon*.

CAPE. A headland, a more or less pointed piece of land jutting out into the sea.

CAPRICORN, TROPIC OF. See *Tropic of Capricorn*.

CAPTURE, RIVER. See *River Capture*.

CARBON DIOXIDE. A gas composed of one part of carbon combined with two parts of oxygen, and normally occupying .03 per cent of the atmosphere.

CARBONIFEROUS PERIOD. The Coal Age: that part of the *Palaeozoic Era* when coal was extensively formed. Thick layers of partially decayed swamp vegetation, covering coastal lowlands, had become buried under marine deposits when the coastal lands sank. More swamps were formed when the water grew shallower, and the process was repeated. As the deposits became compressed and hardened, the vegetable matter formed coal. The coal thus occurs in seams separated by *Sedimentary Rocks*, such as limestone and sandstone, from the marine deposits.

CARDINAL POINTS. The four main directions or points of the compass: north, south, east, and west.

CARSE. In Scotland, the low alluvial land bordering a river near to its mouth, e.g. Carse of Gowrie.

CARTOGRAPHY. The art of drawing maps and charts.

CASCADE. A small *Waterfall*, or a series of falls resembling steps.

CASH CROPS. Crops which are produced for sale, and not for consumption by the farmer and his family. *Plantation* crops are of this kind. See *Subsistence Crops*.

CASTE. An exclusive and hereditary social group which a person involved in the system is allotted, and to which he or she belongs throughout life. The caste system is particularly observed by the Hindus; it confers many privileges on the members of the higher castes, but condemns those of the lower castes, formerly known as the 'untouchables', to the most menial labour, such as scavenging, for life. There are now signs, however, that the rigidity of the system is gradually being broken down.

CATARACT. A great *Waterfall* or series of falls, or sometimes, as on the River Nile, *Rapids*.

CATCHMENT BASIN or DRAINAGE AREA. The region which drains all the rain water that falls on it, apart from that removed by evaporation, into a river or stream, which then carries the water to the sea or to a lake; its boundary is defined by the ridge beyond which water flows in the opposite direction – away from the basin.

CAVE. A hollow space worn out of a portion of the earth's crust. A sea-cave may be produced by the action of the waves, and also by boulders and pebbles being thrown against a cliff by the sea. It may be formed too, by the contraction and expansion of the air in a rock fissure as the waves advance and retreat; the air is compressed by the rise of each wave, and allowed to expand as the wave falls, so that in time the roof and walls of the cave may be broken up and the cave enlarged, even though the waves themselves have not actually come into contact with the rock.

Inland caves are often formed in a limestone region, where water containing carbon dioxide dissolves out underground channels, and enlarges them in places to the dimensions of caves, usually with a stream flowing through them.

CELESTIAL EQUATOR. The imaginary circle formed by the intersection of a plane through the centre of the earth perpendicular to its axis and the *Celestial Sphere*; it thus corresponds in the celestial sphere to the terrestrial *Equator* on the earth.

CELESTIAL SPHERE. A sphere of infinite radius, having its centre at some point within the solar system, as, for instance, at the centre of the earth, on to which all members of the solar system may be projected.

CENOZOIC ERA. See *Cainozoic Era.*

CENTIGRADE SCALE. The temperature scale used over most of the world, i.e. in non-English-speaking countries. It is the simplest scale devised, the two 'fixed points', the Freezing Point and the Boiling Point of water, being respectively 0° and 100°. To convert a temperature on this scale to one on the *Fahrenheit Scale*, it should be multiplied by 1.8, and then 32 should be added: $F.=(1.8 \times C.)+32.$

CENTROSPHERE. See *Barysphere.*

CHAIN. A mountain system consisting of a collection of more or less parallel ranges, and possibly including plateaux, provided that the general longitudinal arrangement is maintained.

CHALK. A soft white or greyish type of *Limestone* which sometimes consists largely of the calcareous remains of small marine organisms and fragments of shells; in its purest form it may contain as much as 99 per cent of calcium carbonate.

CHALYBEATE. Containing iron, e.g. a *Mineral Spring.*

CHANNEL. (1) A relatively narrow stretch of sea between two land masses, and connecting two more extensive areas of sea.
(2) A river bed.
(3) The deep, navigable part of a bay, harbour, etc.

CHAPARRAL. The low, dense *Scrub*, principally shrubs and bushes, which is characteristic of regions having a Mediterranean climate: usually applied to that part of California, U.S.A., which enjoys this climate. It corresponds to the *Maquis* of the Mediterranean area in Europe.

CHART. (1) A map of any type. See *Weather Chart.*
(2) A map to facilitate navigation at sea by indicating noteworthy features, such as the depth of the ocean.
(3) A graph.

CHERNOZEM. See *Black Earth.*

CHILI. The hot, dry, southerly *Sirocco* wind of Tunisia, N. Africa.

CHIMNEY. A narrow cleft in a rock, especially in the Lake District, England, which a climber can ascend.

CHINA CLAY or KAOLIN. A whitish clay formed by the decomposition of the *Feldspar* in *Granite* and extensively used in the manufacture of china or porcelain.

CHINOOK. The warm, dry, *Föhn* type of wind experienced along the eastern side of the Rocky Mountains in Canada and the U.S.A. It normally blows on the southern side of a depression which is moving eastwards across the continent, and is south-westerly, though its direction is considerably modified by local topography; it is thus commonest during winter and spring. Being dried and warmed by descent from the Rockies, the chinook raises the atmospheric temperature, sometimes 30° to 40° F. in fifteen minutes, providing a great contrast to the anticyclonic cold, melts and dries up the winter snow, and so makes grazing possible almost throughout the winter. It is thus of major economic importance, especially in the pastoral regions, from southern Colorado, U.S.A., as far north as the lower Mackenzie River, in Canada. Strong and frequent chinooks mean that the winter is mild, and pastures are available practically without interruption; absence of chinooks means a severe winter, and probably heavy losses of livestock.

CHRONOMETER. An extremely accurate timekeeper, such as is used to determine *Longitude* at sea; in the latter case, it is poised so as to remain horizontal, whatever the position of the ship.

CINDER CONE. See *Ash Cone*.

CINDERS, VOLCANIC. See *Lapilli*.

CIRCUMDENUDATION or CIRCUMEROSION, MOUNTAINS OF. Mountains which remain when less resistant rock masses have been worn away. See *Relict Mountains*.

CIRQUE or CORRIE. A deep, rounded hollow with steep sides, formed through erosion by snow and ice, and thus characteristic of regions which have been glaciated. In regions where glaciers still exist, *Névé* often lies in the cirques, is reinforced by snow from the heights above, and feeds the glacier. In regions where glaciation took place in the distant past, numerous lakes often fill the cirques, and enhance the beauty of the mountain scenery. In Wales the term *Cwm* is often used.

CIRROCUMULUS. A type of high *Cloud*, made of ice crystals, and consisting of small flakes or globular masses, in groups or lines, sometimes known as a '*Mackerel Sky*'.

CIRROSTRATUS. A uniform, thin, milky veil of high *Cloud*, which does not blur the outline of the sun or moon, but produces *Halos*.

CIRRUS. A type of high *Cloud*, consisting of detached pieces, delicate and feathery in appearance.

CIVIL TWILIGHT. See *Twilight*.

CLAY. An exceptionally fine-grained substance, very retentive of moisture, often becoming plastic when mixed with water. Of the various types of clay, which differ according to texture and also origin but are all grouped together as *Sedimentary Rocks*, many contain large amounts of silicates of alumina. See also *Boulder Clay*.

CLEARING. A piece of land cleared for cultivation, especially the primeval forest.

CLEAVAGE. The splitting of a rock, which has been submitted to great pressure, into thin sheets or slabs. The direction of cleavage is occasionally parallel to the stratification of the rocks, but usually is inclined or at right-angles to it; like jointing, it is due to planes of weakness in the rock. See *Joint*. Roofing slate provides an example of a rock in which cleavage is well developed.

CLIFF. A high and extremely steep rock face, approaching the vertical, either inland or along a coastline.

CLIMATE. The average weather conditions of a place or region throughout the seasons. It is governed by latitude, position relative to continents and oceans, and local geographical conditions. Broadly speaking, the interiors and the eastern parts of the great continents have a continental climate, with small rainfall, low humidity, and a great range of temperatures, both diurnal and seasonal, while oceanic islands and the western parts of continents have a heavier rainfall, higher humidity, and more uniform temperatures; there are numerous exceptions, however, and local climates are still further modified by altitude, proximity of mountains, etc.

Near to the equator, climate is almost synonymous with *Weather*, as there is so little variation in the latter. But between the Tropics and the Poles, especially in the region of the *Westerlies*, the weather is often so variable that the climate can scarcely be described in concise terms.

CLIMATIC REGION. One of the main areas into which the earth is divided according to *Climate*. All geographers do not employ the same nomenclature, but the four principal climatic regions which are usually recognized are the tropical, sub-tropical, intermediate or temperate, and polar. These may then be sub-divided into other regions. In the tropical region, for instance, are the equatorial (see *Equatorial Forest*), *Monsoon*, Sudan type (see *Savanna*), and dry or hot desert regions (see *Desert Climate*). In the sub-tropical region are the dry sub-tropical or *Mediterranean Climate* and the wet sub-tropical or *Cotton Belt Climate*. In the intermediate or temperate region are various sub-divisions of the *Maritime Climate* and the *Continental Climate*. Finally, there is the polar or Arctic climate (see *Ice-Cap Climate* and *Tundra*). Where the normal characteristics of any of these regions are masked by altitude, a *Mountain Climate* may be said to prevail.

CLIMATOLOGY. The science which treats of the various climates of the earth, and their influence on the natural environment.

CLIMATOTHERAPY. The treatment of disease through suitable climatic environment, often, but not always, found in recognized health resorts. As climate is subject to seasonal variations, the required environment may have to be sought in different localities at different periods of the year.

CLIMOGRAPH. A graphical representation of the differentiation between various types of climate. Mean monthly values of the *Wet Bulb Temperature* as ordinates are plotted against the *Relative Humidity* values as abscissae, and a closed, twelve-sided polygon, the climograph, is obtained. This reveals the type of climate at a glance: a climograph showing wet bulb temperatures and relative humidities which are all high, for instance, depicts a constantly hot, damp climate. See *Hythergraph*.

CLINOMETER. The instrument used to determine the *Dip* of a rock stratum, the slope of an embankment, etc.

CLINT. See *Grike*.

CLOUD. A mass of small water drops or ice crystals, formed by the *Condensation* of the water vapour in the atmosphere, usually at a considerable height above the earth's surface: the water vapour being created by evaporation of surface water – oceans, lakes, rivers, etc. Although clouds assume an almost infinite variety of forms, two main types are recognized, according to their shape and mode of formation: cumuliform or 'heap' clouds – clouds of great vertical depth – and stratiform or 'layer' clouds. Meteorological classification of clouds, internationally agreed upon, is more specific, however, and describes ten principal forms, which for convenience are separated according to their approximate height above the earth's surface as low, medium, and high clouds. About 8,000 ft is usually taken as the upper limit of low clouds, medium clouds occurring within the range of 8,000 to 15,000 ft, and high clouds above 15,000 ft; these heights are applicable to temperate latitudes, for in the tropics they tend to be greater. The ten forms above-mentioned are:

(1) *Low Clouds: Stratocumulus, Nimbostratus, Cumulus, Cumulonimbus, Stratus.*

(2) *Medium clouds: Altocumulus, Altostratus.*

(3) *High Clouds: Cirrus, Cirrostratus, Cirrocumulus.*

There are, in addition, certain sub-divisions of these.

CLOUDBURST. An abnormally heavy downpour of rain, usually associated with a *Thunderstorm*; in temperate regions it is necessarily short-lived, for the available supply of water vapour is soon exhausted by it. During this short time, however, it may cause considerable damage, tearing up the ground and transforming gullies into raging torrents. Cloudbursts, though possible anywhere that rain falls, are most frequent in mountainous districts. There they are sometimes caused by the sudden stopping of the upward movement of heated air as a storm crosses a mountain range; with the cutting off of this supply of rising air, the raindrops and hailstones which had been supported by it fall in much shorter space of time than if the supply had been maintained.

CLOUDINESS. The state of the sky with respect to clouds, or, in meteorological and climatological terms, the amount of sky covered by cloud, irrespective of the type; it is estimated visually, and is usually expressed in eighths of sky covered. By this scale, o represents a cloudless sky, and 8 a sky entirely covered by cloud. A map of mean annual cloudiness shows two belts of high cloud amount, corresponding to the equatorial and circumpolar low pressure belts; areas of very low cloud amount are shown, of course, in the hot deserts. See *Isoneph*.

CLUSE. A narrow gorge or *Transverse Valley* cut through a mountain ridge; the term is used chiefly of the Jura Mountains.

COALING STATION. See *Fuelling Station*.

COAST. That part of the land which borders the sea or other extensive tract of water, and so comes under the direct influence of the waves.

COASTAL PLAIN. A plain which borders the sea coast, and extends from the sea to the nearest elevated land. It is sometimes formed through denudation by the sea, the beach being later raised by earth movement to form a plain, frequently known as a *Raised Beach*, or by deposition of solid matter at their mouths by rivers.

COASTLINE or SHORE-LINE. The coastal outline of the land, which includes bays, but crosses narrow inlets and river mouths.

COL. (1) A high pass through a range of mountains, usually formed by two streams on opposite sides of the ridge cutting back towards one another and so lowering the *Water-Parting*.

(2) The region situated between two *Depressions* which face one another, and also between two *Anticyclones* which face one another, the depressions and anticyclones being placed alternately; the nearly uniform pressure in a col is therefore higher than in the two depressions which it separates, and lower than in the two anticyclones. Sometimes on a weather chart only the two anticyclones appear, the col being an area of relatively low pressure between them – like a pass between two mountains; sometimes, too, only the two depressions may be shown. On the edge of the col, the winds circulate in different directions round the anticyclones and depressions, and within the col are light and variable. The weather associated with a col in summer is sometimes fine, but there is a pronounced tendency to the development of thunderstorms; in winter, conditions are liable to be dull or foggy.

COLD DESERT. See *Ice Sheet*.

COLD FRONT. The boundary line at the earth's surface between a mass of advancing cold air and a mass of warm air, beneath which the cold air pushes like a wedge. The frontal surface rises at a steeper angle than in the case of the *Warm Front*. The passage of a cold front through a place is normally marked by a rise of atmospheric pressure, a fall of temperature, a veer of wind, a heavy shower, and sometimes a line squall, perhaps with thunder.

COLD POLE. A name frequently applied to Verkhoyansk, in eastern Siberia, where excessively low temperatures have been reached; the mean midwinter temperature is −50°C., and the lowest reading ever taken on the earth's surface, −70°C., has been recorded. This is due, in the first place, to the development during winter of an intense anticyclone over Siberia, and, in these quiet weather conditions, the accumulation of a vast pool of abnormally cold air. The village of Verkhoyansk, situated almost on the Arctic Circle, lies at the bottom of a steep-walled valley carved into the plateau. The already cold air on the plateau is still further chilled by contact with the snow, which loses heat by radiation very rapidly in the clear, dry atmosphere – a loss which continues throughout the long polar night; this chilled air then sinks into the valley, and gives the excessively low temperatures at Verkhoyansk.

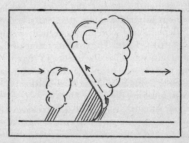

*Upward movement of warm air at a Cold Front (broken arrow),
and direction of movement of the front.*

COLD WAVE. A burst of cold air, usually of polar origin, often experienced after the passage of a depression, behind the *Cold Front*. Such cold waves are specially prevalent in North America and Siberia, for during winter a reservoir or pool of cold air has accumulated over the great continental land masses; in the southern hemisphere the cold waves are comparatively mild, for the smaller land masses there do not permit such a vast accumulation of cold, anticyclonic air. See *Southerly Burster, Pampero.* Cold waves are always menacing to crops.

As employed by the United States Weather Bureau, the term cold wave signifies a specific fall of temperature over twenty-four hours to a minimum below a certain temperature, the amount of fall and the minimum limit varying according to the season and the locality.

COLONY. A human settlement formed by a body of people in a territory far from their native land, usually in an undeveloped and sparsely inhabited country; as used in the political sense (not, for instance, as in speaking of the French colony in London), the term signifies the country so settled – and in this sense the colony is subject to the country from which the colonists hailed.

COMBE, COOMB, COOMBE. In southern England, a hollow or short valley in the side of a hill, especially in the chalk downs.

COMET. A heavenly body revolving round the sun in an eccentric orbit, consisting of a solid nucleus and a luminous, gaseous tail, and small in comparison with a planet.

COMMON. A tract of land which belongs to the local community as a whole, and is open to common use.

COMPASS, MAGNETIC. An instrument consisting of a magnetized needle, which, by being balanced on a fine point, is free to rotate in a horizontal plane; the needle always swings to such a position that one end points towards the North Magnetic Pole – *not* the North Geographical Pole. See *Magnetism, Terrestrial*. In the mariner's compass, the magnetic needle is fixed to a circular card, likewise free to rotate in the horizontal plane, which is graduated into thirty-two divisions of $11\frac{1}{4}°$ each – the thirty-two points of the compass.

COMPOSITE LANDSCAPE. A landscape which exhibits topographic features that have developed in more than one *Cycle of Erosion*.

CONDENSATION. The process by which a substance changes from the vapour to the liquid state. Clouds, for instance, are formed by the condensation of water vapour in the atmosphere; air containing water vapour rises from the earth's surface, expands, and is cooled till the water vapour condenses into drops of water. When larger drops have collected and begin to fall as rain, they may evaporate again before reaching the earth's surface. The opposite process to condensation, in fact, is *Evaporation*.

CONDOMINIUM. A territory governed jointly by two or more countries: e.g. the New Hebrides, governed jointly by Great Britain and France.

CONE, ASH or CINDER. See *Ash Cone*.

CONFIGURATION. The form of a part of the earth's surface with respect to both its horizontal outline and its elevation.

CONFLUENCE. The point at which one stream flows into another, or where two streams converge and unite.

CONFLUENT. A stream flowing into another of roughly equal size.

CONGLOMERATE. A rock composed of rounded pebbles cemented together in a matrix of finer material (often sandy); it is thus consolidated *Gravel* or *Shingle*. See also *Breccia*.

CONICAL PROJECTION. The type of *Map Projection* produced when the globe is imagined as being surrounded by a paper cone, the apex above the pole; the map is projected on to the cone, and the latter then unrolled. Straight lines radiating from the apex of the cone represent meridians, concentric circles represent parallels of latitude. Only along the parallel where the cone touched the globe are the distances on the two equal, so that the projection is accurate only along this one parallel, and is known as the 'one-standard parallel' projection. It is therefore used only for sections of the globe, the distortion being inconsiderable for a small area.

Errors are reduced in several modified conical projections, e.g. the 'two-standard parallel' projection, in which distances are made accurate along two parallels. The latter are selected, one towards the top, the other towards the bottom of the map, and both are made the correct length and the true distance apart. This scale is correct along all meridians and along the selected standard parallels, and the projection is thus superior to the 'one-standard parallel' projection for all maps having a substantial extent of latitude.

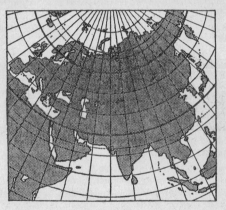

Conical Projection with two standard parallels.

CONIFEROUS FOREST.* A forest of evergreen coniferous or cone-bearing trees carrying needle-shaped leaves. From such forests is obtained the valuable softwood timber of commerce. The most important and extensive coniferous forests of the world stretch in a huge belt across northern Canada and northern Eurasia, with 'tongues' of forests extending southwards on the higher land, e.g. along the Rocky Mountains, and 'islands' of forest land in other mountainous areas farther south, e.g. in the Alps. These coniferous forests form the home of many valuable fur-bearing animals.

CONJUNCTION. (1) The apparent proximity to each other of two heavenly bodies.

(2) The position of two heavenly bodies when they are in the same or nearly the same direction as viewed from the earth.

CONSEQUENT RIVER. A river which flows down the initial slope of land, i.e. in the direction of the *Dip*, and is thus a necessary consequence of that slope. See *Obsequent River*, *Subsequent River*.

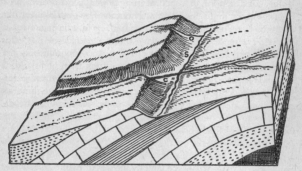

Diagram to illustrate Consequent (C), Obsequent (O),
and Subsequent (S) rivers.

CONTINENT. (1) One of the larger, unbroken masses of land into which the earth's surface is divided – Europe, Asia, Africa, North and South America, Australia, and Antarctica.

(2) In a colloquial sense, the mainland, applied particularly to Europe by the people of the British Isles.

CONTINENTAL CLIMATE. The type of climate experienced in the interior of the great continents, especially in the intermediate or 'temperate' zone of the northern hemisphere. Its influence also extends to eastern coasts of the continents and neighbouring islands, which are exposed in winter to the prevailing winds from the interior. It is a climate characterized by extremes of temperature, with maxima and minima occurring soon after the summer and winter *Solstices* respectively. In addition to the great range of temperatures experienced, both diurnally and seasonally, other features of the climate are a relatively small rainfall and low humidities. See *Maritime Climate*.

CONTINENTAL DRIFT. The supposed horizontal displacement of portions of the original continent which comprised the entire land mass of the world to form the present-day continents; it was notably described in Wegener's Hypothesis. See *Displacement Theory*.

CONTINENTAL PLATFORM. The approximately level part of the earth's crust which is raised above the depressions in which the oceans lie, and includes both the lower land areas of the continents themselves and also the *Continental Shelf* which borders the continents.

CONTINENTAL SHELF. The sea bed, bordering the continents, which is covered by shallow water, in general 100 fathoms or less in depth; it thus takes the form of a shelf or ledge sloping gently downwards from the coast, and is approximately outlined by the *Isobath* of 100 fathoms. It varies considerably in width, reaching in some places 100 miles or more. Where it is widest, the angle of slope is usually least, and may be less than 1°; where the coast is mountainous, it is usually narrow, and there is a quick transition from high land to deep water. Beyond the Continental Shelf there is a drop in the sea bed to a depth of 1,000 fathoms or more, so that the shelf has a relatively steep edge. As the land slopes continuously downwards on to the Continental Shelf, with no change of angle at the water-line, it is widely held that the edge of the shelf represents the former boundary of the continent; the shelf may have been formed by a rise in level of the sea or a fall in level of the land, or by the *Denudation* of the fringe of the land by the sea, or by the *Deposition of* solid materials beneath the water by rivers, etc.

CONTINENTAL SLOPE or BATHYAL ZONE. The steep slope which descends from the edge of the *Continental Shelf* to the deep ocean bed.

CONTOUR. A line drawn on a map to join all places at the same height above sea level; a number of contours therefore depict on a flat map the relief of the land over the area covered. The intervals between contours may represent height differences of 50 ft as on an Ordnance Survey map, or of as much as several thousand feet, as on a small-scale map of a large area in an atlas. On physical maps, the areas between adjacent contours are often shown in different colours: usually lowlands are coloured green, higher lands brown, the shades deepening progressively with height, while the highest land is often coloured red and finally white. (See Fig. on p. 43.) Similar lines showing the depths of the sea bed are called *Isobaths*, areas between them being coloured in varying shades of blue.

CONURBATION. An area which is occupied by a mass of streets, factories, dwellings, etc., possibly enclosing small isolated rural localities, and formed by the growth of several neighbouring and formerly separate towns.

CONVECTION. The transmission of heat from one part of a liquid or gas to another by the movement of the particles themselves. When the lower portion of a mass of fluid is heated, it expands, its density is reduced, and it rises, carrying its heat with it – to be replaced by cool fluid which in its turn is heated. A familiar example of convection is the upward movement of air which has been heated by contact with the earth's surface; this air is said to rise in a convection current.

CONVECTIONAL RAIN. Rain which is caused by the process of *Convection* in the atmosphere. When the surface layers of the atmosphere are heated, the moisture-laden air rises in a convection current, and in rising is cooled till the *Dew Point* is reached, and its water vapour condenses and forms cloud; the convection current is sometimes so strong that the cloud attains great vertical depth, and its water content becomes so considerable that heavy rain is deposited. Thus the thundery rain of a summer afternoon in temperate regions is typical convectional rain. See *Rainfall*.

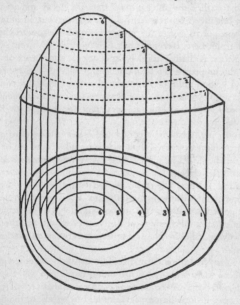

Diagram to illustrate the construction of Contours on a map.

COPPICE or COPSE. A small wood, or plantation of trees, which are periodically cut before growing into large timber.

COPSE. See *Coppice*.

COOMB, COOMBE. See *Combe*.

CORAL ISLAND. A *Coral Reef* situated far from any other kind of land.

It may consist simply of a mound of sand resting on a flat coral reef, in which case it may be several miles long and quite broad, but will have only slight elevation. When there has been an uplift of the land, however, the reef itself may be above sea level, and the island may reach a considerable altitude. By various agencies a coral island may receive seeds, which become established and at length cover it with a rich vegetation.

CORAL POLYP. A small marine creature, constructed like a sea anemone, which has a hard skeleton formed by the solidification of the base and side-walls of the body, and can only live in shallow tropical seas. The skeleton is made of calcium carbonate extracted from the sea water in which the polyp lives, and the solidification of its base fixes it to the surface to which it is attached. When the coral polyp dies, the softer parts of the body decay and are washed away, but the skeleton is left behind. The increase in numbers of the coral polyps causes the masses of coral to grow to enormous size, forming *Coral Reefs* and *Coral Islands*.

CORAL REEF. A chain of rocks lying at or near the surface of the sea, and built up principally by immense numbers of *Coral Polyps*; both on and behind the reef, fragments of shells, coral, and coral sand are piled up by wind and wave, and new land is thus formed. Three more or less distinct kinds of coral reef are recognized: fringing reefs, barrier reefs, and *Atolls*.

A *fringing reef*, formed by corals and other calcareous organisms growing on the bed of the shallow seas round a continent or island, lies near the shore; its surface comprises a rough, uneven platform at about the level of low water, and its outer edge slopes downwards into the sea. The corals grow outwards from the shore to a depth of about 30 fathoms, and upwards to low water level, so that the edge of the platform lies at approximately the 30-fathom line. As the outer corals grow most freely, they may reach the surface first, leaving a shallow channel or lagoon between the platform and the land.

A *barrier reef* lies at some distance from the shore, its outer side rises from deep water where no reef-building corals live, and the lagoon is wide and deep. There are two possible explanations of its formation: either its foundations were not laid by corals, or the depth of the sea has increased since the corals built them; both may be true. See *Atoll*.

CORDILLERA. A group of mountain systems, each system probably comprising several ranges; applied especially to the great Andean mountain mass of South America, it may be applied too, for instance, to the group of four mountain systems in western U.S.A., from and including the Rocky Mountains to the Pacific coast.

CORN BELT. The belt of land in the United States, situated south and south-west of the Great Lakes, where maize – known to the Americans as corn – is principally grown. The states which lead in corn production in this area are Iowa, Illinois, and Nebraska.

CORN BELT CLIMATE. The type of climate experienced in the *Corn Belt* of the United States, and in other regions similarly situated on the continents of Europe and Asia which together produce most of the world's corn crop: e.g. the plains of the Danube basin, and northern China. The climate may be regarded as the long-summer type of the cool intermediate climate, or as the long-summer subdivision of the humid *Continental Climate*.

CORONA. (1) The fringe of radiant light surrounding the sun which is plainly visible round the dark rim of the moon during a total eclipse. (2) A number of coloured rings round the sun or moon, comprising a bluish inner region and a reddish outer ring. Usually this, the aureole, is all that appears; sometimes, however, there is a complete corona, the aureole being surrounded by a series of coloured rings, from violet on the inside through blue, green, yellow, to red. Thus the corona is distinguished from the *Halo* by the fact that its colour sequence is opposite; it is produced by the diffraction of light by water drops, while the halo is due to refraction.

CORRASION. The mechanical *Erosion* caused by loose solid material during its *Transportation*: e.g. that effected by the solid material carried down by a river, in wearing away fragments from its bed and its banks. Corrasion of the bed, thereby causing a deepening of the channel, is called *vertical corrasion*, while corrasion of the banks is called *lateral corrasion*. Although a stream of clear water will not affect solid rock, it will cause considerable corrasion if the bed and banks consist of loose material. Most of the corrasion effected by a river, however, is due to the pebbles and sand that it drives along its bed. Another example of corrasion is the wearing away of the surface of desert rocks by wind-borne sand.

CORRIE. See *Cirque*.

CORROSION. The wearing away of rocks by chemical action, e.g. by solution, as distinct from *Corrasion*, which is a mechanical process.

COSEISMIC LINES. See *Homoseismal Lines*.

COTTON BELT. The belt of land in the south-eastern United States which produces more of the world's cotton crop than any other area; the rich black soils of Texas, Mississippi, and Alabama have the highest output, but most of the state of Florida lies outside the Cotton Belt.

COTTON BELT CLIMATE. A term used by some geographers to signify the type of climate experienced in the American *Cotton Belt*, which forms part of the *Sub-Tropical Region*. The climate, in fact, may be regarded as one of the two sub-divisions of the sub-tropical climate—the humid or wet sub-tropical climate. It differs from the dry sub-tropical or *Mediterranean Climate* in being usually located on the eastern sides of the continents, in having a heavier rainfall, and in having its rainfall well distributed throughout the year or concentrated in the warm season. The climate is characteristic of southern China as well as the American Cotton Belt.

COUNTER-TRADES. See *Anti-Trades*.

COURSE. See *River*.

COVE. A small creek, inlet, or bay.

CRAG. A rough, steep rock or point of rock. In high mountain regions crags are often formed by the action of frost in *Weathering*.

CRAG-AND-TAIL. A hill or crag having a steep, often precipitous face on one side, and sloping downwards comparatively gently on the other side. When the crag-and-tail has been formed glacially, there is a hollow on one side of the crag and a long 'tail' of broken rock material on the other side. This is due to the fact that the crag or mass of rock, lying in the bed of a glacier, has obstructed its flow; the ice has then worn away the rock on the upstream side, and deposited the waste on the downstream side.

CRATER.* The funnel-shaped hollow at the top of the cone of a *Volcano*. The bottom of the funnel opens into the channel or pipe through which the erupted material finds its way to the surface. The term is also sometimes applied to the hollow caused by the fall of a *Meteorite* on to the earth's surface.

CRATER-LAKE. A sheet of water collected in the *Crater* of an extinct volcano. See *Caldera*.

CREEK. (1) A narrow coastal inlet.
(2) A small stream, tributary, or branch of a river.

CREOLE. A person born in the West Indies, but not of the aboriginal race; also applied to such persons born in certain areas of the American mainland (particularly Central America) and some European colonies. The term is thus used sometimes of the descendants of European settlers, sometimes of Negroes.

CREVASSE.* A deep, vertical crack in a glacier. It may be formed when the slope down which the glacier is moving steepens, for there the glacier begins to move more quickly. When the slope becomes gentler again, the crevasse tends to close up, but it rarely makes the surface of the ice smooth once more; the sun has melted some of the ice while the crevasse has been open, and so widened its upper part, and the two opposite faces do not fit when the crevasse closes. A crevasse is also formed when the glacier has to turn a sharp corner. Both these kinds of crevasses are transverse to the glacier, and, since it moves more rapidly in the middle than at the sides, they become curved outwards in the direction of movement. More or less longitudinal crevasses may be formed when the valley widens and the glacier spreads; oblique crevasses may also develop, owing, for example, to the difference between the velocity in the middle and at the sides of the glacier. The term is also applied to a wide crack in the raised bank or levee of a river or canal.

CROFT. A small farm-holding in the Highlands or Islands of Scotland, consisting of a piece of land used primarily for cultivation, usually adjoining the house. When combined with fishing and possibly other occupations, the croft is reckoned to supply all essential family needs in food, fuel, and clothing, though its average size, excluding the common hill pastures over which the few cattle and sheep can roam, is only about five acres.

CROMLECH. A prehistoric structure, consisting of a large flat stone resting on upright stones; in Great Britain the term is practically synonymous with *Dolmen*, while in France it is applied to a circle of small *Menhirs*.

CRUSTAL MOVEMENTS. The movements of the outer parts of the solid *Lithosphere*, as manifested by *Raised Beaches*, *Earthquakes*, etc.

CUESTA (Spanish). A ridge, or belt of hilly land, formed on gently dipping rock strata (e.g. on a *Coastal Plain*) from the more durable layers, which resist denudation better than the weaker layers, and are thus left behind as uplands; it has a gentle *Dip* slope on one side, and a relatively steep *Scarp* on the other.

CUMULONIMBUS. A heavy, dark *Cloud* of great vertical depth, towering upwards in huge, voluminous masses, the tops being 15,000 ft or more above the base, often spreading out aloft in the form of an anvil. Below the base, very low, ragged clouds frequently trail. It is the typical thunderstorm cloud, and usually gives showers of rain, snow, or hail.

CUMULUS. A *Cloud* of considerable vertical development, though less so than in the *Cumulonimbus*. The upper parts are dome-shaped, and have 'cauliflower' heads, while the base is practically horizontal. Two types are generally recognized: the fair-weather cumulus, small, white, detached puffs typical of fair weather, and heavy cumulus, which has much greater depth, and often develops later into cumulonimbus.

CURRENT. See *Ocean Current*.

CUT-OFF. See *Ox-bow Lake*.

CWM. In Wales, a term used principally for a *Cirque*, but occasionally for other types of valley.

CYCLE OF EROSION. The series of changes through which *Erosion* causes a newly uplifted land surface to pass, from youth through maturity to old age. In youth, streams occupy steep-sided V-shaped valleys. *Corrasion* is rapid, and such features as rapids, waterfalls, and lakes are often apparent. In maturity, or middle age, when erosion has proceeded farther, the valleys are broad and gentler in slope, and rivers have begun to meander. In old age, valleys are very broad, rivers are sluggish, and the region becomes a *Peneplain*.

CYCLONE. A region of low atmospheric pressure, of which there are two types. The first, characteristic of temperate latitudes, is now usually referred to as a *Depression*; the second, a much more violent phenomenon, though in general covering a smaller area, and typical of the tropics, is usually called a *Tropical Cyclone*. The two types are similar in the fact that their winds in the northern hemisphere circulate in an anti-clockwise direction, and in the southern hemisphere in a clockwise direction.

CYCLONIC RAIN. Rain associated with the passage of a cyclone or *Depression*, and caused by a warm, moist air mass moving upwards over colder, heavier air. See *Rainfall*.

CYLINDRICAL PROJECTION. The type of *Map Projection* produced if a paper cylinder is imagined as surrounding the globe, and, after all relevant points on the latter are projected on to the former, the paper is then opened out. On this kind of projection the poles cannot be truly depicted, as their points have been extended to occupy the whole width of the map, and, as a result, all territories lying near to the poles are much exaggerated in size. This is a serious disadvantage, and with the one exception of *Mercator's Projection* cylindrical projections are not in common use.

D

DALE. A wide, open valley, mainly in northern England and southern Scotland.

DAM. An obstruction formed by a glacier or by other natural agency across a river or stream so as to produce a lake, or a similar man-made obstruction built across a river in order to control the flow of water. In general the purpose of constructing a dam is to provide a supply of water, particularly for irrigation, and to manufacture electricity cheaply by utilizing the head of water thus artificially produced.

DATUM LEVEL. The zero with reference to which the altitudes of land surfaces are determined. See *Mean Sea Level*.

DAWN. The faint light which illuminates the various regions of the earth before sunrise; the time when light appears in the sky; the interval during which the atmosphere is illuminated before sunrise. See *Twilight*.

DAY. See *Sidereal Day* and *Solar Day, Mean*.

DEAD VALLEY. The term used, particularly by French geographers, for a *Dry Valley*.

DÉBÂCLE. The breaking up during spring or summer of the ice formed on rivers during the winter, chiefly applied to the great rivers of the U.S.S.R. and North America. In general it takes place at successively later dates with increasing distance from the equator: in the southern U.S.S.R. about mid-March, but in extreme northern Siberia not until June. The débâcle has a duration of two to six weeks, and during this time the rivers often overflow their banks and inundate the surrounding country.

DÉBRIS. See *Detritus*.

DECIDUOUS FOREST. A forest consisting of trees which lose their leaves at some season of the year. In the case of the monsoon forests, such as those of India and Burma, the trees shed their leaves during the hot season in order to protect themselves against excessive loss of moisture by evaporation. In the case of the deciduous forests of the cool intermediate lands, such as those of north-west Europe, the trees shed their leaves during the autumn (appropriately called the 'fall' in America because then the leaves are falling), in order to protect themselves against the cold and frost of winter. From the deciduous forests is obtained much of the valuable hardwood timber of commerce: from the monsoon forests such extremely hard wood as teak, from the cool intermediate forests such wood as oak, elm, and beech.

DECLINATION. The angular distance of a heavenly body from the *Celestial Equator*, measured on a meridian passing through the body.

DECLINATION, MAGNETIC or VARIATION, MAGNETIC. The angle at any point on the earth's surface between the magnetic meridian (the direction in which a freely pivoted compass needle points) and the geographical meridian (the true geographical north). At most places on the earth's surface the compass points either west or east of true north, and the declination could be expressed, for instance, as 10° W. See *Agonic Line*. Magnetic declination is subject to a regular diurnal variation, to irregular, comparatively short-period changes, and to secular change. See *Magnetism, Terrestrial*.

DECLINATION OF THE SUN. The angular distance of the sun either north or south of the equator. At the summer *Solstice*, about June 21, the *north* declination of the sun is thus 23½°, and about December 22 the sun's *south* declination is 23½°. A sailor, by observing the sun's zenith distance in degrees, and allowing for the declination (found from the Nautical Almanac), can thus find his latitude at noon on any day; if he is north of the Tropic of Cancer, for instance, and the sun is north of the equator (i.e. in summer), he will calculate his latitude by adding the sun's declination to its zenith distance.

DEEP. One of the deepest parts of the ocean, forming a depression in the sea floor of limited area, and having relatively steep sides. The deeps are not usually found in the middle of the oceans, but towards the margins, generally where volcanoes are still active and earthquakes common. Most of the deeps, in fact, occur as narrow, trough-like depressions around the shores of the Pacific; among them are the Tuscarora Deep off Japan and the Atacama Deep along the coast of South America. There are others along the edges of some of the plateaux which rise from the *Deep-Sea Plain* – one extending from the Tonga Islands almost to New Zealand. A *Sounding* of 5,763 fathoms was recorded north-east of Mindanao in the Philippines, and in 1960 the U.S. bathyscaphe 'Trieste' descended to 5,967 fathoms in the Marianas Trench. In the Atlantic, where there are fewer deeps, a sounding of 4,561 fathoms was recorded in the Blake Deep, north of Puerto Rico, and the only other deep with soundings above 4,000 fathoms is the Romanche Deep, near the equator on the western side, where a depth of 4,030 fathoms was recorded.

DEEP-SEA PLAIN. The extensive and fairly level area, about 2,000 to 3,000 fathoms below the surface, which forms the majority of the ocean floor. In places it descends to the *Deeps*.

DEFILE. A term rather loosely used, but usually applied to a gorge, ravine, or narrow pass.

DEFLATION. The lifting and conveyance of particles of sand and dust by the wind.

DEFORESTATION. The process of clearing forests.

DEGRADATION. A process which tends to wear down the land surface; it is usually applied to a river, and involves the deepening of its valley by the river. When the load of loose, solid material entering a stretch of river, for instance, is less than the amount it can carry there, the river wears away from its bed a greater amount than it deposits. The term is also applied to a glacier. See *Aggradation*.

DEGREE. (1) The unit of temperature, varying according to the scale employed, e.g. *Centigrade, Fahrenheit*.

(2) The unit of measurement of *Latitude* or *Longitude*, usually employed to indicate position on the earth's surface. As a degree is geometrically equal to 1/360 of a circle, the distance between the equator and the pole, which is one quarter of the earth's circumference, is taken to be equal to 90° of latitude; a degree of latitude is thus equal to 1/360 of the earth's circumference. Similarly, a degree of longitude is also equal to 1/360 of the earth's circumference *at the equator*, but decreases in length in passing from the

equator to either of the poles, becoming zero at the poles. Each
degree of latitude or longitude is divided into 60 minutes ('), and
each minute into 60 seconds (").

DELL. A small wooded valley; a dale.

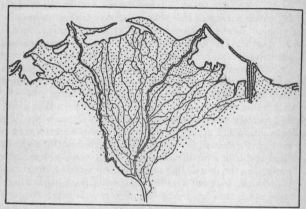

Sketch map of the Delta of the Nile.

DELTA. The fan-shaped alluvial tract formed at the mouth of a river,
when it deposits more solid material there than can be removed by
tidal or other currents: the name being originally applied by the
Greeks to the alluvial tract thus produced by the Nile, owing to its
resemblance in shape to the fourth letter of their alphabet. As material is deposited, it becomes easier for the river to divide and flow
to each side, each new branch forms new banks, then divides and
sub-divides, and the deposit grows outwards in the shape of a fan or
triangle, which finally becomes covered with a network of channels; the apex of the triangle is called the head of the delta. A small
delta is sometimes built where a river enters a lake, or at the confluence of two rivers, especially where a swift stream heavily laden
with sediment joins a slow river.

Many of the great deltas of the world are extensively cultivated,
and some support dense agricultural populations. When the delta
streams make sudden great changes of course, as they are liable to
do, disastrous floods may affect these populations. In the Hwang-
ho flood of 1887, at least a million people were drowned and many
villages destroyed. In addition to the Nile and the Hwang-ho, other
well-known deltas include those of the Mississippi and the Ganges-
Brahmaputra.

DEMOISELLE (French). See *Earth Pillar*.

DENDRITIC DRAINAGE. The type of drainage produced when a *Consequent River* receives a number of tributary streams, which in turn are fed by smaller tributaries, the whole forming a dendritic or tree-like pattern.

DENSITY OF POPULATION. The average number of inhabitants living within a specified unit of area – usually, in English-speaking countries, the square mile – in a certain region.

DENUDATION. The wearing away of the land by various natural agencies: the sun, the wind, the rain, frost, running water, moving ice, and the sea. The heat of the sun causes the rocks to expand, crack, and break up; the wind carries loose particles, and with sand helps to wear away rocks; the rain loosens and carries away soil; frost freezes water in the cracks of rocks, causes the cracks to widen, and the rock to break; rivers and streams wear away the land, especially in their upper courses, where their slopes are steep; moving ice in the form of glaciers also wears away the land; the sea along the coasts wears away the rocks. In addition, water, especially if it contains carbon dioxide, has a strong solvent action on parts of the earth's crust. Rock salt is readily and completely soluble, limestone more slowly and less completely soluble; most rocks, however, are only partially dissolved. The process of denudation may thus be classified as *Weathering, Transportation*, with its accompanying *Corrasion*, and solution. Denudation is one of the two major processes responsible for earth sculpture, or the change in form of the earth's surface, the other being *Deposition*.

DEPENDENCY. A territory which is governed by a state from which it is remote.

DEPOSITION. The laying down of solid material which has been carried from a distant part of the earth's crust by some natural agency. Rivers, for instance, deposit mud and pebbles; the winds carry much sand and dust, which are deposited when they die down, and fall either on the land or into the sea; glaciers carry along boulders and stones, which they deposit where they end owing to the melting of the ice; the sea deposits sand and pebbles. Deposition is one of the two major processes of earth sculpture, the other being *Denudation*.

DEPRESSION. A region where the atmospheric pressure is low relative to that of its surroundings, sometimes known as a low or a cyclone. On the *Weather Chart* it is represented by a number of closed *Isobars*, more or less oval or circular in shape, the isobar of lowest pressure being the nearest the centre of the depression. Both the size and the velocity of a depression are very variable; it may be 100 miles to over 2,000 miles across, it may be practically stationary or moving at about 600 or 700 miles per day. A 'deep' depression is one in which the pressure is very much lower at the centre than at the edges, a 'shallow' depression one in which these pressures are not very different from each other. The winds of a depression, which in the northern hemisphere circulate round the centre in an anti-clockwise direction, in the southern hemisphere in a clockwise direction, are generally much stronger than those of an *Anticyclone*, the *Pressure Gradient* being steeper.

According to modern meteorological theory, the depressions which are so prominent a feature of the weather of temperate regions originate where warm tropical air meets cold polar air, the former ascending over the latter with the formation of frontal surfaces. See *Front, Cold Front, Warm Front*. Unsettled weather is characteristic of depressions, and the rainfall which accompanies them is due to the ascent of the warm air over the cold air, and the formation of extensive cloud systems. See *Cyclonic Rain*. With depressions are often associated *Secondary Depressions*.

Arrangement of Isobars and surface winds in a Depression (N. Hemisphere).

DESERT. An almost barren tract of land in which the precipitation is so scanty or so spasmodic that it will not adequately support vegetation. A desert in which absolutely nothing grows, however, is uncommon; it may be extremely poor grassland or extremely poor scrub. A rock desert is one in which the rock has been exposed owing to the strong erosive action of the wind, with only slight deposition. In a stony desert, the rock surface has been broken up by temperature changes, and the ground covered by a layer of fragments, from which the smaller particles have been carried away by the wind. A sandy desert usually has an undulating surface of *Dunes* with intervening hollows.

The deserts of the world are often sub-divided into tropical or hot deserts and mid-latitude deserts; of the former, probably the best known are the Sahara and the Arabian Desert, of the latter the deserts of Gobi and Turkestan. Those areas of high latitude or high altitude whose lack of vegetation is due to low temperatures are sometimes called 'cold deserts'. See *Desert Climate*.

DESERT CLIMATE. The type of climate of which the salient feature is aridity, or lack of regular precipitation; it may thus occur within a wide range of latitudes, the two principal climatic divisions being the tropical or hot desert and the mid-latitude desert. The *Isohyet* of 10 in. may be taken as a rough guide to the limit of the desert climate, though in the far north of Canada and Siberia a rainfall of 10 in. is actually sufficient to promote forest growth. Character of rainfall, in fact, is as important as amount: in some desert areas there may be a rainfall of over 20 in. during the year, but it falls in violent thundery showers, and is mostly lost by *Run-Off* and evaporation. Associated with the aridity of the desert climate are such characteristics as amount of sunshine, range of temperatures, and relief of land – all of which combine to distinguish it from other climates, in whatever latitude it is situated. Its most significant feature is its unproductiveness, its inability to support settled, independent communities.

One of the main causes of the aridity of the desert climate is distance from marine influence, so that it occurs principally in continental interiors. Another is the system of *Planetary Winds*, with which areas of permanent or seasonal high pressure in the *Horse Latitudes* are associated; high atmospheric pressure is unfavourable to precipitation, and the greatest areas of desert climate in the world are thus situated in these areas of high pressure – the Sahara, Arabia, central Australia, the Kalahari and the Atacama Deserts, etc.

DESERT PAVEMENT. A relatively smooth, mosaic-like area in a desert region, consisting of pebbles closely packed together after the removal of finer material.

DETRITUS or DÉBRIS. Fragments of material removed by disintegration and other processes from the surfaces of rocks.

DEW. The moisture deposited on the earth's surface, or on objects near to the earth's surface such as blades of grass, when nocturnal radiation from the earth has cooled the lower layers of the atmosphere below the *Dew Point*, and the water vapour present has condensed into drops. The greater the amount of water vapour in the air, and the lower the temperature of the ground, the more abundant the dew. For dew to form, the air must be calm or nearly so, in order that it may remain long enough in contact with the ground to be cooled to the *Dew Point* – unless that temperature has already been practically reached. A warm day favours evaporation, increasing the amount of water vapour in the air, and lack of clouds favours radiation at night; thus calm weather and a clear sky provide the best conditions for production of dew. See *Hoar Frost*.

DEW POINT. The temperature at which the atmosphere, being cooled, becomes saturated with water vapour, and by *Condensation* the latter is deposited as drops of *Dew*.

DEW POND. An artificially constructed pond, lined with cement or clay, situated at high level on chalk downs, e.g. in southern England, which retains its water for unusually long periods, even during drought, and is therefore useful for watering cattle. The name suggests that the water supply is maintained by dew, but available evidence is strongly against this theory.

DIASTROPHISM. The disturbance·and dislocation of the earth's crust, including its bending, folding, and breaking, which have resulted in the major inequalities of its surface.

DIATHERMANCY. The thermal permeability, or the ability to transmit radiant heat, of a body such as the atmosphere.

DIATOM OOZE. One of the *Oozes* which cover a large proportion of the ocean bed. It is of vegetable origin, consisting of the siliceous parts of microscopic diatom plants or algae which mainly inhabit the cold waters of the earth; they have shells of silica, from which the ooze is formed. There is a broad belt of diatom ooze extending round the earth in the Southern Ocean.

DIKE. See *Dyke*.

DINGLE. A small, narrow, well-wooded valley.

DIP. (1) The maximum slope of an inclined stratum of rock, measured from the horizontal, and stated in terms of the angle in degrees and the direction as a point of the compass, e.g. 5° NW.

(2) The angle between the direction of the earth's magnetic field and the horizontal at any point on the earth's surface.

DISPLACEMENT THEORY. The theory that the continents are parts of a once larger land mass, and that this mass fractured, parts of it drifting away to form separate continental islands. Evidence exists in the fact that the shores of opposite continents fit together fairly well, like the pieces of a jig-saw puzzle: e.g. the western coast of Africa and the eastern coast of South America. See *Continental Drift*.

DISSECTED PLATEAU. A *Plateau* into which a number of valleys have been carved by erosion; its origin as a plateau is patent, however, when the tops of the mountains and ridges are seen to be level against the skyline, showing that they once formed part of a continuous surface.

DISTRIBUTARY. A branch or outlet which leaves a main river and does not rejoin it, carrying its water to the sea or a lake. It is usually applied to the numerous channels into which a river divides on its *Delta*, or on an *Alluvial Fan*.

DIURNAL RANGE. The amount of variation between the maximum and minimum of any element, such as air temperature, during 24 hours.

DIVIDE. See *Watershed*.

DOAB. Chiefly on the Indo-Gangetic Plain of India, the alluvial tract of land between the two adjacent rivers, e.g. Ganges and Jumna.

DOCTOR, THE. A name given locally to a number of different winds in various parts of the world, on account of their moderating effect on the otherwise unpleasant and unhealthy weather conditions; the *Harmattan* of West Africa and the regular summer sea-breeze of the coast of Western Australia are often so-called.

DOLDRUMS. The equatorial belt of low atmospheric pressure where the NE. and SE. *Trade Winds* converge on and meet each other, producing calms and light surface winds and a strong upward movement of air. Thus, in spite of the calms and light winds, the doldrums are characterized by turbulent and stormy weather, with heavy rains, thunderstorms, and squalls: on account of their lack of wind, they used to be avoided as far as possible by sailing-ships, for the latter were often becalmed for several days. The Doldrums are variable both in position and extent, usually moving northwards and southwards with the sun; their movement, however, is much less than that of the sun, being about 5° on each side of the mean position, with a lag of one or two months behind the sun. Along coasts in the Doldrums, the alternating *Land Breeze* and *Sea Breeze* are important.

DOLINE or DOLINA. A closed hollow in a *Karst Region*, formed by the solution of the limestone near the surface, and subsequent subsidence; it is often rounded or elliptical in shape, and sometimes has a *Sink Hole* into which surface water flows.

DOLMEN. A prehistoric structure consisting of a large, unhewn stone resting on several upright stones; in Great Britain the term is practically synonymous with *Cromlech*.

DOLOMITE. A semi-transparent crystalline mineral consisting of the double carbonates of calcium and magnesium.

DOME. A curved stratum of rock in which the *Dip* is in all directions away from a central point rather than an axis, as in an *Anticline*.

DOMINION. One of the self-governing states within the Commonwealth of Nations – Canada, Australia, New Zealand, India, Ceylon, Pakistan, Malaya, Ghana, Nigeria.

DONGA. In South Africa, a steep-sided gully or dry watercourse, similar to the *Wadi* or *Nullah*.

DOWN. A tract of open, treeless, hilly land, especially applied to the chalk hills of southern England, sparsely covered with soil and mainly used for pasturing sheep.

DOWNLAND. A term sometimes applied to the mid-latitude or temperate *Grasslands* of Australia and New Zealand.

DRAINAGE AREA. See *Catchment Basin*.

DREIKANTER. See *Ventifact*.

DRIFT. (1) A slow movement of surface water at sea, on a lake, etc., caused by the wind; the similar movement of sand, etc.

(2) Any deposit on the earth's surface which has been transported, e.g. *Loess* (by the wind), *Boulder Clay* (by ice); the term is often used specifically of boulder clay.

(3) In South Africa, an important ford on a river.

DRIFT-ICE. Detached portions of ice-floes or icebergs carried by currents into the open sea, beyond the limit of the pack-ice.

DRIZZLE. Rainfall in which the water drops are extremely small, like a fine spray.

DROUGHT. An extended period of dry weather; extreme dryness due to lack of rain. A drought is liable to cause most distress in a region on the boundary between one which has an abundant or at least reliable rainfall and one in which the rainfall is adequate in most years but from time to time fails – as in some of the densely populated parts of India and China. In the less populous parts of the interior of Australia, North and South America, the losses due to drought are mainly confined to livestock. Certain definitions of types of drought have been adopted. See *Absolute Drought, Partial Drought, Dry Spell*.

DROWNED VALLEY or SUBMERGED VALLEY. A valley which has been submerged by the advance of the sea or a lake, owing to the sinking of the land. See *Ria*.

DRUMLIN. An elongated hill or ridge of *Boulder Clay*, usually oval and shaped like half an egg; it occurs in a previously glaciated region, the long axis lying parallel to the direction of flow of the ice, with the thick, steep end to the north. The exact manner of formation of drumlins is somewhat uncertain, but they are probably fragments of ground *Moraine* which were compressed by ice movements into their characteristic shape. Wherever they are found, they occur in large numbers. Some of the best-known groups of drumlins are those situated in Northern Ireland (e.g. Co. Down), northern England (e.g. Eden valley), and northern U.S.A. (near Madison,

Wisconsin, and south of Lake Ontario in central New York State).

DRY FARMING. A method of farming without irrigation in an area of limited rainfall, the land being treated so as to conserve the moisture it contains; the term is usually confined to the raising of crops. Specially adapted plants that will best utilize the available moisture are cultivated. Other features of dry farming are the removal of weeds that would take up some of the moisture, and the preparation of a mulch to resist capillarity and protect the moist earth from the heat of the sun. A common practice is the formation of a dry, powdery surface soil by frequent use of the plough, the harrow, and other implements, and to this procedure is sometimes added the artificial production of a hard, impervious sub-surface layer of soil. Even stones may be used as a mulch, as in the drier parts of the Mediterranean region. In Siberia, where melting snow provides valuable moisture for the spring-sown crops, the soil is ploughed in the autumn; this conserves the snow in the furrows, and prevents it from being swept from the ground by the strong winds of winter, and thus being lost to the soil. In the drier regions of India, U.S.S.R., U.S.A., Canada, and Australia, dry farming methods have long been employed.

DRY SPELL. In the British Isles, a period of at least fifteen consecutive days, none of which has received as much as .04 in. of rain. The definition has not been internationally accepted, however, and a dry spell is often defined, not by any specific set of climatological data, but by its effects on plant life. See *Wet Spell*.

DRY VALLEY. A valley in which the water of a stream has disappeared, or almost disappeared, possibly owing to its diversion, or to the extreme permeability of the soil. The name used by French geographers for such a valley, a 'dead valley', is better, for some water may continue to flow through it.

DUNE.* A hill or ridge of sand formed, either in a desert or along the sea coast, by transportation by the wind. The sand particles are carried along by the wind and piled into a heap, which gradually increases in size till it becomes a small hill; the dune is often commenced where an obstacle of some kind impedes the free movement of the wind, and the sand is heaped against this until it covers it and falls over on the other side. The windward side, against which the sand is blown, normally has a gentle slope, and the leeward side, formed by the falling sand, is steep, except when the dunes consist of long, narrow ridges running in the direction of the prevalent wind, when both the sides and also the leeward end are steep.

By the action of the wind, the shape and size of a dune are always changing. When the winds are variable, it has no fixed shape, and its movements are irregular. When the winds are fairly constant in direction, however, the sand on the windward slopes is blown over the top and falls over on the leeward side, and the dune thus travels slowly forwards. On the margin of a sandy desert, this movement of the dunes has sometimes caused cities to be buried, as in Egypt and Syria, or has devastated once fertile land. In the *Landes* of SW. France, such movement was checked by sowing plants and later coniferous trees. Often, however, it is only the smaller dunes which move: when the occasional rain of the hot desert does fall, much of the water is conserved in the deeper parts of the large dune, evaporation being restricted to the surface layers; this moistness of the lower layers thus helps to fix the position of the dune. See *Barkhan*, *Seif Dune*.

DUST. Solid matter consisting of minute particles, smaller than sand particles, and occurring everywhere in the atmosphere; it is often carried immense distances by the wind, and is constantly being deposited on the earth's surface. The sources of dust are various: in and near to industrial areas, the smoke from factory and domestic chimneys charges the atmosphere with particles of carbon and other substances; in the deserts, dust is raised from the ground by the wind (see *Duststorm*), while volcanic dust enters the air during an eruption.

DUST BOWL. A region which is subject to severe *Drought* and *Duststorms*, originally and still mainly applied to the Dust Bowl of the western United States.

DUST COUNTER. An instrument for counting the number of dust particles present in a given volume of air.

DUST DEVIL. A local whirl of dust, usually not more than a few yards in diameter, in which the particles are swept round and round the centre, and are lifted to considerable heights – sometimes to 2,000 or 3,000 ft above the earth's surface. It is caused by excessive local heating by the sun in an arid region, a strong convection current being formed. Dust devils usually move over the desert at from 5 to 15 miles per hour, though speeds of over 30 miles per hour are sometimes reached, and as many as five or six of them may often be counted by the observer. Being so limited in size, they are harmless even to aviation – unlike the *Duststorm*.

DUSTSTORM.* A storm in which a thick mass of dust obscures the atmosphere, and reduces visibility very considerably – sometimes practically to zero. It causes acute discomfort on the surface, and is a danger to aviation. It is caused by a turbulent wind blowing over an arid, dusty surface, and its approach is usually heralded by a wall of dust which extends upwards for several thousand feet – sometimes to over 10,000 ft. In advance of the storm, the wind is light and variable, but increases in strength as the storm approaches.

Duststorms are frequent in such arid desert regions as the plains of Iraq, North Africa, and NW. India, and are especially liable to occur in thundery weather; the atmosphere is so dry that the rain associated with the thunderstorm often evaporates before reaching the ground, but if it does not, the dust is washed down with it, and the duststorm is thus much shortened in duration and restricted in scope. It should be distinguished from a *Sandstorm*.

DUST-WELL. A hollow, containing dust, in the surface of a glacier; it is formed when a patch of dust, blown on to the ice, absorbs heat from the sun more rapidly than does the ice itself, and causes the latter to melt, a depression being thus created.

DYKE or DIKE. (1) A vertical or highly inclined sheet of igneous rock, formed when molten rock material or *Magma* from the interior of the earth has forced its way towards the surface through a cleft, or by melting a passage for itself, and has there cooled and solidified. The dyke may change its direction, or branch away into smaller dykes, or give rise to *Sills*; its thickness may vary from a fraction of an inch to hundreds of feet.

(2) A ditch.

(3) A bank of earth, stones, etc., constructed to prevent low-lying land from being inundated by the sea, a river, etc.

E

EAGRE. A tidal wave or *Bore*.

EARTH. The fifth in size of the eight major planets, and the third in distance from the sun. The solid outer crust of the earth, known as the *Lithosphere*, is partially covered by an extensive area of water, known as the *Hydrosphere*, and around the earth is a gaseous envelope known as the *Atmosphere*. Although the earth is often regarded as being a sphere, it is actually an oblate spheroid, being slightly flattened at the poles; the polar diameter is about 27 miles shorter than the equatorial diameter, the latter being 7,926 miles in length.

EARTH PILLAR, DEMOISELLE (French), or HOODOO. A tall column of earth, often 20 to 30 ft high, capped by a large boulder. The boulder has originally lain on the soil; when most of the soft surface material surrounding it has been gradually worn away by rain, it has protected the earth beneath it, and has remained perched on a long pillar of this earth. Earth pillars occur frequently in mountain valleys, and probably the best-known examples are those of the Tyrol. (See Fig. on p. 61.)

EARTHQUAKE. A movement or tremor of the earth's crust which

originates naturally and below the surface. It sometimes causes a permanent change of level at the surface, but often the damage done by the shaking provides the only lasting visible effect. It may be produced by a volcanic explosion; earthquakes, in fact, are common in most volcanic areas, and often precede or accompany eruptions. It is more likely, however, to be of *Tectonic* origin, and probably due to the existence of a *Fault*.

Diagram of an Earth Pillar, typical of those seen in the Tyrol.

At least three distinct sets of waves are set up by an earthquake, and at a considerable distance from the place of origin these are felt separately; close to the place of origin, however, they all reach an observer at approximately the same time. As these waves pass a place, the ground may be felt to rock, and buildings sway backwards and forwards. Maximum damage is not done always at the *Epicentre*, where the movement is up and down, but at places where the wave reaches the surface obliquely yet which are still close enough to the origin for it to have lost little of its force. A large earthquake is usually followed by a series of other shocks. An earthquake which originates below or near to the sea causes great disturbance of the water, and sometimes large waves emanate from it and travel a considerable distance; on occasion these waves have caused a greater loss of life, by flooding coastal regions, than the earthquake itself. See *Tsunami*.

In the main earthquake regions, many of which have active volcanoes in their midst, an earthquake of some kind takes place practically every day; on Hawaii, for instance, hundreds of minor shocks are recorded during the year. There are extensive areas on the earth's surface, however, where earthquakes are very rare. The three great regions where earthquakes have taken place more or less frequently are (1) the west coast area of North and South America, (2) a belt across southern Europe and southern Asia, and (3) a belt in the Pacific Ocean which includes Japan, the Philippines, and most of the East Indies. Of the thousands of earthquakes which are recorded annually, only a hundred or so cause damage.

EARTHQUAKE PERIOD. The period during which a region is subject to earthquake shocks without any extended respite.

EBB-TIDE. The receding or falling tide, i.e. after high tide and before low tide. See *Tides*.

ECLIPSE, LUNAR. The obscuration of the light of the moon when the earth, passing between sun and moon, casts her shadow on the latter. It can thus occur only at the time of full moon, when the moon is in opposition. It does not occur at every full moon, however, for the moon's orbit is not in the plane of the *Ecliptic*, but is inclined at about 5° to it. If the moon is completely obscured, there is said to be a *total eclipse*, if partially, a *partial eclipse*. Lunar eclipses are far less frequent than solar eclipses, but in a given area many more of them are visible than of solar eclipses. This is because at every place where the moon is above the horizon at the time of the eclipse, the eclipse is visible, whereas a solar eclipse is visible only over a limited area. In a lunar eclipse the *Penumbra* is scarcely visible. A total lunar eclipse may last as much as two hours. See *Eclipse, Solar*.

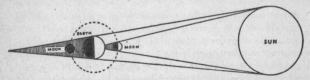

Simplified diagram to illustrate Solar Eclipse and Lunar Eclipse.

ECLIPSE, SOLAR. The obscuration of the light of the sun when the moon, passing between earth and sun, casts her shadow on the former. It can thus occur only at the new moon, when the moon is in conjunction, but it does not occur at every new moon because of the inclination of the moon's orbit to the plane of the *Ecliptic*. If the sun is completely obscured, there is said to be a *total eclipse*, if partially, a *partial eclipse*; when the moon's shadow does not even reach the earth, an *annular eclipse* is given – a ring of light is visible round the circular disc of the moon. Solar eclipses are much more frequent than lunar eclipses, but they can be seen only over a very limited area: the cone of the moon's shadow becomes so narrowed across the space between moon and earth that a total eclipse is observed along a narrow belt. The average diameter of the shadow of a total eclipse, in fact, is 90 miles. Unlike the lunar eclipse, however, its *Penumbra* is visible, and a partial eclipse is seen over a much wider area. On account of the restricted area from which any total solar eclipse is visible, at any single place the phenomenon is rare, and expeditions are organized to the locality from which it can be observed; its duration is only a few minutes.

ECLIPTIC. The apparent track of the sun throughout the year, as a result of the motion of the earth round it, being a great circle on the *Celestial Sphere*. The *plane of the ecliptic* is the plane passing through this path, and is coincident with the plane of the earth's *Orbit*; it thus meets the celestial sphere in the ecliptic.

ECOLOGY. The science which treats of organisms in relation to their environment; it is frequently sub-divided into human ecology, animal ecology, plant ecology, and bio-ecology, the last-named dealing with the interrelationships between animal life and plant life. Ecology lies on the frontiers of so many other subjects, including various branches of geography, that its limits have not yet been precisely defined.

EDAPHIC. Relating to the soil.

ELEVATION. (1) The raising of a portion of the earth's crust in relation to its surroundings. See *Subsidence*.

(2) The altitude or angular height of an object such as a heavenly body above the horizon.

ENCLAVE. An outlying territory belonging to one country which lies wholly within the territory of another country. The term is sometimes loosely used to denote an area which is *almost* enclosed in this way: e.g. one which is bordered by the sea on one side.

ENTRENCHED or INCISED MEANDER. Part of an old *Meander* which has become deepened by rejuvenation. See *Rejuvenated River*.

ENTREPÔT. A place or district which acts as an intermediary centre for trade between foreign countries: that is, for the receipt of goods from one part of the world and their distribution to another part; or a place where goods are temporarily stored. Two well-known examples of seaports outstanding in entrepôt trade are Port Said and Hong Kong.

EPEIROGENESIS. Continent-building: large-scale changes in the level of the earth's crust, in which the surfaces of continents have been simply elevated or depressed, with little or no folding. The formation of a *Plateau* provides an example. See *Orogenesis*.

EPICENTRE or EPICENTRUM. The point on the earth's surface

which is vertically above the point of origin or *Seismic Focus* of an earthquake. Here the earthquake vibrations first reach the surface, and then seem to spread outwards. If the time of arrival of the earthquake is observed for several different places, the position of the epicentre may usually be determined. See *Homoseismal Lines*.

EPICONTINENTAL SEA. A shallow sea which rests on a *Continental Shelf*.

EPIPHYTE. A plant which grows on another plant but is not parasitic, obtaining its essential foods from the atmosphere. Epiphytes are typical of the *Equatorial Forest*, and the orchids provide the best-known examples.

EQUAL-AREA PROJECTION. See *Homolographic Projection*.

EQUATOR. The imaginary circle, lying midway between the poles, formed at the surface of the earth by a plane drawn through the centre perpendicular to its *Axis*; as its centre is also the centre of the earth, it is a *Great Circle*. It is also the longest circumference of the earth.

EQUATOR, CELESTIAL. See *Celestial Equator*.

EQUATORIAL FOREST or TROPICAL RAIN FOREST. The hot, wet, evergreen forest of the equatorial region, where rainfall is very heavy and where there is no dry season; it extends in parts into the typical monsoon areas. Owing to the extreme heat and moisture, the growth is dense and luxuriant. Many of the trees grow to tremendous heights, and in the struggle to reach the sunlight *Lianas* climb up other trees; *Epiphytes* are also found abundantly in the equatorial forest. Individual species of trees are very scattered, but among them are such valuable tropical hardwoods as mahogany and ebony; the wild rubber tree is also a native of the equatorial forest. Typical equatorial forest regions include the basins of the Amazon and the Congo, and much of the East Indies.

EQUINOX. The time of the year when the sun appears vertically over-head at noon at the equator. It is thus the time when the illuminated half of the earth just includes the two poles, and when all places on the earth have equal day and night – twelve hours each. The sun rises exactly in the east and sets exactly in the west everywhere. There are two equinoxes each year, one, falling about March 21, being called the *spring* or *vernal equinox*, and the other, falling about September 20, the *autumnal equinox*.

EQUIPLUVE. A line joining all places which have equal *Pluviometric Coefficients*. A map showing such lines thus illustrates the raininess of a particular month or other period over a given area.

ERG. In the Sahara, that part of the desert whose surface is covered with sand *Dunes*. See *Reg, Hammada*.

EROSION.* The wearing away of the land surface by various natural agencies, the most important being those consisting of water – the sea, rivers, and rain. Ice, in the form of glaciers, frost, and melting snow, also assists in the process of erosion.

ERRATIC BLOCK or ERRATIC. A boulder which has been transported from its source by a glacier, sometimes over a considerable distance, and has been left stranded when the ice melted; it is thus, often, of a different type from neighbouring rocks. In some areas, such as parts of South Dakota, Minnesota, and other northern states of the U.S.A., erratic blocks have been so numerous as to impede agriculture.

ERUPTION, VOLCANIC. The forcing of materials in solid, liquid, or gaseous form from the interior of the earth on to its surface by a *Volcano*.

ESCARPMENT or SCARP. An inland cliff or steep slope, formed by the erosion of inclined strata of hard rocks, or possibly as a direct result of a *Fault*. See *Cuesta*.

ESKER. A long, narrow ridge of sand and gravel which was once the bed of a stream flowing beneath or in the ice of a glacier, and was left behind when the ice melted. See *Ås*.

ESTUARY. The mouth of a river where tidal effects are evident, and where fresh water and sea water mix. If the land consists of soft rocks, the funnel-like shape of the estuary may be due to the scouring action of the river and the tide, but in most cases it is due to subsidence of coastal lowland. The river valley has been flooded by the sea, and the tides have maintained rather than actually created the shape of the estuary. Many estuaries, such as those of the Thames, the Elbe, the Plate, are the sites of important seaports.

ÉTANG (French). A shallow pool or lake lying among sand dunes and gradually becoming filled with silt: applied especially to Languedoc, along the Mediterranean coast of France.

ETESIAN WINDS or MELTEMI. The strong, constant, northerly winds of summer experienced in the eastern Mediterranean, blowing round the great *Trough* of low atmospheric pressure which extends westwards from NW. India. From about mid-May to mid-October they regularly blow at 10 to 30 miles per hour, sometimes rising to over 40 miles per hour. During the day they increase in strength, as the insolation over the hot land to the south and east temporarily raises the *Pressure Gradient*, then at night they often decline and die away. On land they sometimes carry considerable dust. Despite the prevailing heat, however, their dryness causes them to be quite refreshing. Their strength and dryness are such that orchards often have to be protected from them by a row of cypresses set on the northern side.

ETHNOGRAPHY. The subject which deals with the regional distribution of the races of mankind.

ETHNOLOGY. The science which treats of mankind as distributed over the earth in racial units.

EUSTATIC MOVEMENT. A large-scale rise or fall of sea level. Such a movement may be caused, for instance, by the development and decay of *Ice Sheets*.

EVAPORATION. The process by which a substance changes from the liquid to the vapour state. Evaporation of surface water by the heat of the sun, from the oceans, lakes, rivers, etc., is the cause of the water vapour in the atmosphere. As the atmosphere is seldom completely saturated, evaporation goes on almost all the time, the rate of evaporation being dependent on the amount of water vapour already in the atmosphere, the nature of the water surface, and the wind. As heat is the main cause of evaporation, the latter is in general greatest in tropical regions. The opposite process is *Condensation*.

EXFOLIATION. A *Weathering* process which consists in the peeling off of thin layers of rock from the surface. In hot deserts it is caused by the heating of the rocks by day and their cooling by night, leading to alternate expansion and contraction. The corners of rock masses especially are broken off, and the surfaces assume a rounded form. The process is often assisted by others, such as the chemical weathering of the outer layers. The term 'onion weathering' is sometimes used for exfoliation.

EXPORTS. Goods which are dispatched from a country in trade, normally because they are surplus to its requirements; from the national standpoint, they may be regarded as goods exchanged for *Imports*. They have usually been produced in the exporting country, but sometimes they may have been purchased abroad and reexported. See *Entrepôt*. See also *Invisible Exports*.

EXTENSIVE CULTIVATION. A system of farming by which the cultivator expends a limited amount of labour and capital on a relatively large area. See *Intensive Cultivation*.

EXTRUSIVE ROCKS. Rocks formed by solidification of *Magma* above the earth's surface, e.g. volcanic *Lava*; they are thus *Igneous Rocks*. See *Intrusive Rocks*.

F

FAHRENHEIT SCALE. The temperature scale used in most English-speaking countries in the world. The 'fixed points', the Freezing Point and Boiling Point of water, are 32° and 212° respectively.

Thus, to convert a temperature on the Fahrenheit scale to one on the *Centigrade Scale*, 32 must first be subtracted from the former, then the result divided by 1.8: $C = (F-32) \div 1.8$.

FALL. The term used in the United States for autumn, signifying the season of 'the fall of the leaf'.

FALL LINE. The line joining the waterfalls on a number of approximately parallel rivers: especially applied to the line running along the sudden increase of slope – where the waterfalls thus occur – from the Appalachian Mountains to the Atlantic coastal plain in the eastern U.S.A. The Fall Line marks the point where each river leaves the uplands for the lowlands, and thus the limit of its navigability. For this reason, and because water power is available there, many important industrial centres have become established along the line, e.g. Philadelphia, Baltimore, Richmond.

FATA MORGANA. A type of *Mirage* in which an object appears to be elongated vertically, owing to the fact that it is observed through several horizontal layers of air of different refractive indices.

FATHOM. The unit of length, equal to 6 ft, used mainly in determining the depth of the sea.

FATHOMETER. An instrument used to determine the depth of the ocean, making use of the known velocity of sound waves through water; a sound wave is transmitted from the surface of the sea to the ocean bed, and by obtaining the time interval between its transmission and its reception again at the surface, after reflection from the ocean bed, the depth of the latter can be easily calculated.

FAULT. A fracture in the earth's crust along which movement has taken place, and where the rock strata on the two sides therefore do not match. Although the movement is usually more or less vertical,

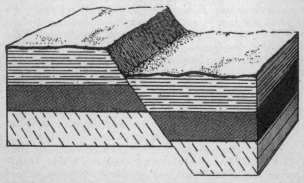

Diagram to illustrate the arrangement of the rock strata in a Fault.

a fault may take place in any direction, including the horizontal. The change of level on one side compared with the other, known as the *throw*, varies from a fraction of an inch to thousands of feet; the amount of lateral displacement is known as the *heave*. The side of a fault where the strata have moved relatively downwards is known as the *downthrow* side, the other being the *upthrow* side. A *normal fault* is one in which the *Fault Plane* is vertical, or is so inclined that the downthrow is on the dip side of the fault plane. See also *Thrust Fault*.

FAULT PLANE. The surface along which a *Fault* has taken place.

FAULT SCARP. A *Scarp* situated where a *Fault* has taken place, being due to the relative downward movement of the strata on the lower side of the fault.

FAUNA. The animal life of a region or of a geological period, corresponding to the term *Flora* for plant life.

FELDSPAR or FELSPAR. A widely distributed group of minerals consisting of silicates of aluminium combined with potassium, sodium, calcium, and barium, and probably comprising almost half of the earth's crust. Although insoluble, feldspar undergoes chemical decomposition, *China Clay* being a product.

FELL. Chiefly in northern England, a bare, uncultivated hill or mountain.

FELLAH (pl. FELLAHIN). In Arabic-speaking countries, especially Egypt, a peasant; in Arabic the term means literally 'a ploughman'.

FEN. A tract of low-lying, marshy land: applied especially to the region round the Wash in eastern England.

FERREL'S LAW. The law which states that a body moving in any direction over the earth's surface will tend to be deflected, owing to the earth's rotation, to the right in the northern hemisphere and to the left in the southern hemisphere; it is applied especially to the movement of large masses of air – the winds. Thus a wind which would blow from N. to S. if the earth were stationary becomes in the northern hemisphere NE., e.g. the NE. Trade Wind.

FERTILIZER. A substance of natural or artificial origin which is added to the soil to increase its productivity, by supplying those chemical elements that are necessary to plant life.

FETCH. At sea, the length of open water across which the wind is blowing, i.e. the distance from the weather shore, which largely determines the height of the waves.

FIORD. A long, narrow inlet into the sea-coast, with more or less steep sides. Owing to the occurrence of fiords along the edges of mountainous regions which were once heavily glaciated, e.g. along the coasts of Norway and Greenland, it is generally held that they were formed when glaciers, making their way to the sea, scooped out deep, trough-like valleys which had been first cut by streams, so that the lower ends were filled by the sea. Usually a fiord is extremely deep, but becomes shallower towards the mouth, possibly because the glacier lost some of its power of erosion as it melted, or because a terminal *Moraine* blocked the end.

Fiords along the coast of southern Norway.

FIRN. See *Névé*.

FIRTH. In Scotland, a long, narrow inlet into the sea coast, usually the lower part of an *Estuary*, but sometimes a *Fiord*; occasionally a *Strait*, e.g. Pentland Firth.

FLOE. See *Ice-Floe*.

FLOOD-PLAIN. A plain, bordering a river, which has been formed from deposits of sediment carried down by the river. When a river rises and overflows its banks, the water spreads over the flood-plain; a layer of sediment is deposited at each flood, so that the flood-plain gradually rises. It is usually highest near the river, at the natural *Levee*. During the formation of a flood-plain, it is often characterized by *Marshes*, meandering streams (see *Meander*), and *Ox-Bow Lakes*. If the drainage system is later diverted for some reason, the flood-plain may be left as a smooth and extremely fertile tract of land.

FLOOD-TIDE. The advancing or rising tide, i.e. after low tide and before high tide. See *Tides*.

FLORA. The plant life of a region or of a geological period, corresponding to the term *Fauna* for animal life.

FOG. A dense mass of small water drops, or smoke or dust particles, in the lower layers of the atmosphere. According to international meteorological practice, the phenomenon is termed a fog only when the obscurity is such that objects at a distance of 1 kilometre from the observer have become invisible. See *Mist, Haze*.

The condensation of water vapour in the lower layers of air, which is normally the cause of fog, is due to the cooling of the air below its *Dew Point*. This cooling may be due to *Radiation* from the earth's surface, and the conditions most likely to produce a fog are thus a very light wind and a clear sky at night. The fog is then most dense shortly after sunrise, and often disperses before the afternoon; in winter, however, it may persist for several days. A fog may also be formed by the drift of air over a cold surface (see *Advection Fog*), or by the mixture of one *Air Mass* with a cooler air mass. Land fogs occur chiefly in autumn and winter, sea fogs chiefly in spring and summer. The depth of a fog is very variable, but is usually below 1,000 ft; at sea it is sometimes so shallow that the masthead of a ship will project above it. A smoke fog, in which the poor visibility is due mainly to smoke particles, occurs in industrial areas where there is a continual emission from factory and domestic chimneys; but the worst type of fog in these areas is produced by a mixture of smoke and water drops (see *Smog*). A dust fog is typical of a desert region.

FOG BOW. A white bow seen opposite to the sun in a fog; it resembles and is produced in the same way as an ordinary *Rainbow*, but owing to the smallness of the water drops, the colours overlap and the bow is white – though the inner edge has a bluish tinge and the outer edge a reddish tinge. Sometimes a second bow is seen inside the first, the colours being reversed.

FÖHN WIND. The warm, dry wind which blows down the leeward slope of a mountain, best known in the valleys of the northern Alps, where the name originated. It occurs when a depression moving to the north of the Alps draws in air from the south. This air ascends the southern slopes, clouds are formed, and there is heavy rain; when it reaches the northern slope, it has lost most of its moisture but is still warm, and it is dynamically heated further as it descends, and so blows down the valleys as a very warm, dry wind – the föhn. It may raise the temperature by 8° to 11° C.; snow is melted, trees, houses, etc., become excessively dry, avalanches may occur. The maximum frequency of the föhn in northern Switzerland is in spring, when it is useful in melting the winter snow from the pastures. In autumn, the season of second greatest frequency, it is useful in ripening the crops, especially the grapes. The föhn blows in the direction of the mountain valleys, but is usually a southerly wind, and is strongest in valleys lying in a north-south direction. A similar type of wind occurs in all mountainous regions which are affected by depressions. See *Chinook, Nor'Wester, Samoon*.

FOLD. A bend in rock strata caused by movements of the earth's crust. When the compression is relatively small, the strata are formed into a series of arches and troughs (see *Anticline* and *Syncline*); when the compression is greater, the folds bulge outwards at the top, and lower down dip inwards from both sides. *Denudation* takes place simultaneously with the folding, so that the outer parts of a fold may be worn away as it is produced.

FOLDED or FOLD MOUNTAINS. Mountains which have been thrown up into a massive *Fold* or ridge by earth movements. Often the mountains form an *Anticline*, and the adjacent valleys are *Synclines*, though where such mountains have been exposed to erosion for a long period there is a tendency towards the reverse, the anticlines to become valleys and the synclines to become hills, for the resistance of the latter to the denuding effect of water is greater than that of the former. Most of the important mountain ranges consist of folded mountains, including the Himalayas, the various Alpine chains, the Pyrenees, and the Apennines.

FORD. The shallow part of a river or other body of water which may be crossed by wading.

FOREST. An extensive area of land covered with trees. See *Coniferous Forest, Deciduous Forest, Equatorial Forest*.

FOSSE. A ditch or moat.

FOSSIL. The remains or the form of a plant or animal which has been buried and preserved for a long period in the rocks of the earth's crust. The only organisms which have been thus preserved are those which were buried in *Sedimentary Rocks* soon after death, or were encased in a material which protected them from decay and the attacks of living organisms. See *Palaeontology*.

FOUCAULT'S PENDULUM. The apparatus by which Foucault demonstrated the rotation of the earth on its *Axis*. A heavy ball was suspended by a fine wire from the dome of the Panthéon, in Paris, so that it could swing freely in any direction, and the pendulum was set in motion; gradually the direction of the swing appeared to change, and, as no external force had been applied to the pendulum, the rotation of the floor, and of the earth itself, was thus shown.

FRAZIL ICE. Surface ice which forms in spicules on rapidly flowing rivers, the movement of the water preventing the ice crystals from forming a solid sheet. The phenomenon has been best observed in the Canadian rivers, and the term is Canadian in origin.

FREE PORT. A port, or a zone within a port, at which goods, either imports or exports, may be loaded and unloaded without payment of customs duty. It is usually an *Entrepôt*. Hong Kong is an example of a free port.

FRESHET. (1) A clear stream.

(2) A river flood caused by heavy rains or by rapidly melting snow.

FRIAGEM. The *Cold Wave* experienced on the tropical *Campos* of Brazil during winter, caused by the development of an anticyclone. In the region of the middle Amazon, for instance, there is sometimes a cool spell during May or June lasting several days. The temperature may fall below 10° C., causing acute discomfort, and rendering the natives extremely vulnerable to colds.

FRIGID ZONE. The regions within the *Arctic Circle* and the *Antarctic Circle*. The sun's rays are always very oblique, so that the weather in general is never warm: in summer it is cool or cold, in winter it is extremely cold.

FRONT. The line of separation at the earth's surface between cold and warm air masses. In general it is produced by the horizontal movement of two such air masses, which have originated from widely separated regions, e.g. tropical and polar air. This brings them into contact with each other, and from the line of the front on the ground a frontal surface, or surface of discontinuity, slopes upwards over the cold air; the warm air, being lighter than the cold air, is continually ascending this frontal surface. Usually a front lies along a *Trough* of low atmospheric pressure. See *Depression, Cold Front, Warm Front.*

FRONTIER. That part of a country which borders on another country, i.e. adjoins the boundary-line between the two countries.

FROST. (1) The particles of frozen moisture formed on the earth's surface when the air temperature has fallen as far as or below 0° C. See *Glazed Frost, Hoar Frost, Rime.*

(2) An air temperature, at or below 0° C., which causes water to freeze; the intensity of the frost, i.e. the number of degrees of frost, is indicated by the number of degrees C. that the air temperature falls below the Freezing Point of water (0° C.). In temperate regions, e.g. the British Isles, a frost may be very local and transitory, owing probably to cooling by radiation, or it may continue for several days over an extensive area.

FRUIT BELT. A strip of land about 30 miles wide extending along the eastern shore of Lake Michigan, U.S.A., where the lake so tempers the climate that grapes and similar fruits can be successfully grown, whereas their cultivation would be impossible in other districts in the same latitude.

FUELLING STATION. A port, or a repository within a port, which supplies fuel in the form of coal or oil to ocean-going vessels which are unable to carry enough in their bunkers for their complete voyage.

FUMAROLE. A hole in the earth's crust from which steam and gases, such as carbon dioxide, are emitted under pressure. It is frequently found in a volcanic region; the steam is produced by the *Magma* beneath the surface, when the volcano itself has probably ceased to erupt. The most striking development of fumaroles in the world is in the Valley of Ten Thousand Smokes, near the volcano Katmai, in Alaska; here an area of several square miles is riddled with fumaroles ejecting steam.

FUNNEL CLOUD. The cloud formed at the core of a *Waterspout* or *Tornado*, sometimes extending downwards as far as the earth's surface.

G

GABBRO. A group of coarse-grained *Igneous Rocks* consisting essentially of a *Feldspar* and one or more ferro-magnesian minerals.

GALE. According to common usage, simply a very high wind, but according to usual meteorological practice a wind of Force 8 or more on the *Beaufort Scale*: i.e. a wind of about 35 miles per hour or more on the surface.

GALL'S PROJECTION. A *Cylindrical Projection* in which the cylinder, instead of touching the globe only at the equator, is sunk into its surface, and coincides with the surface at 45° N. and 45° S. latitude.

GANGUE. The material, of relatively small value, surrounding or accompanying a metallic *Ore* in a lode or vein; it is usually non-metallic, such as quartz.

GARIGUE or GARRIGUE. The scattered, low *Scrub*, interspersed with extensive patches of bare soil, which prevails in the Mediterranean region on land either too poor or too dry for the growth of normal scrub. It is typical of limestone regions. From the standpoint of agriculture, the garigue is almost valueless, being suitable only for feeding goats. Many of the plants which constitute the garigue, such as broom and gorse, have bright but short-lived flowers; others, like lavender and thyme, are aromatic.

GARUA. The thick mist or drizzle experienced in western Peru during the winter, often covering the coastal region like a thick blanket of cloud; it provides almost all of the scanty rainfall of this region.

GAUGE, RAILWAY. The distance between the inner edges of the rails on a railway track. Gauges vary in different countries and are roughly classified as broad, standard, and narrow gauges. The broad gauges usually include those of 5 ft or more in width; the standard gauge, as used in Great Britain, is 4 ft 8½ in. wide; the narrow gauges include the metre gauge, the 3 ft 6 in., and smaller gauges.

GAUGE, RAIN. See *Rain Gauge*.

GEANTICLINE. See *Geoanticline*.

GEEST. The heath-covered sandy region of the glaciated northern lowlands of Germany, lying between the Elbe and the Ems; the soil is infertile and the population sparse.

GEOANTICLINE or GEANTICLINE. An *Anticline* on a large scale, i.e. extending over several miles; its formation is very gradual, and erosion of the crest proceeds simultaneously.

GEODESY. The science of the measurement of the shape and size of the earth, including its weight, density, etc., and also of the surveying of such large portions of the earth's surface that the curvature of the earth has to be considered.

GEOGRAPHICAL MILE. A measure of length equal to one-sixtieth of a degree or one minute (1′) of latitude; it thus varies with latitude, but its approximate value is 6,080 ft. See *Mile, Statute*; *Nautical Mile*.

GEOGRAPHY. The subject which describes the earth's surface – its physical features, climates, vegetation, soils, products, peoples, etc., and their distribution. For its data it has drawn extensively on specialized sciences, such as geology, meteorology, astronomy, anthropology, and biology. A number of sub-divisions of the subject are recognized, including the following: *mathematical geography*, which deals with the shape, size and movements of the earth; *physical geography*, which usually includes a study of *Climate, Natural Vegetation* and *Oceanography*, and is sometimes assumed, rather loosely, to be synonymous with *Physiography*; *human geography*, which is sometimes regarded as synonymous with *Anthropogeography*; *political geography*, which is concerned primarily with the world's political divisions; *economic* or *commercial geography*; and *historical geography*.

GEOID. A term sometimes used to signify the shape of the *Earth*; the latter is often taken to be an oblate spheroid, but, in view of certain variations, the term geoid, which simply means 'earth-shaped body', has been introduced.

GEOLOGY. The science of the composition, structure, and history of the earth. It thus includes the study of the materials of which the earth is made, the forces which act upon these materials and the resulting structures, the distribution of the rocks of the earth's crust, and the history not only of the earth itself but also of the plants and animals which inhabited it throughout the different ages. The subject is often sub-divided into *physical* or *dynamical geology* and *historical geology*; the former includes *Mineralogy, Petrology*, Structural Geology, and *Geomorphology* or *Physiography*, while the latter includes *Stratigraphical Geology* or Stratigraphy, *Palaeontology*, and *Palaeogeography*.

GEOMORPHOLOGY. The study of the physical features of the earth, or the arrangement and form of the earth's crust, and of the relationship between these physical features and the geological structures beneath. The term is sometimes regarded as being synonymous with the older term, *Physiography*.

GEOPHYSICS. The study of the physical processes relating to the structure of the earth, including not only the *Lithosphere* but also the *Hydrosphere* and the *Atmosphere*. Strictly speaking, it signifies the 'physics of the earth'; it is thus a branch of experimental physics, and links the science of physics with that of geology. Such subjects as *Meteorology*, *Hydrology*, and *Seismology* are closely allied to and in some respects are branches of geophysics.

GEOSYNCLINE. A large depression or trough in the earth's crust, or a *Syncline* on a large scale, i.e. extending for distances up to hundreds of miles. A fundamental difference in structure, however, is due to the fact that the geosynclinal trough gradually fills with sediment as its floor subsides, resulting in the formation of an enormous thickness of sedimentary rocks.

GEYSER.* A *Hot Spring* which at regular or irregular intervals throws a jet of hot water and steam into the air; sometimes the jet rises to a height of 100 to 200 ft. It occurs in a volcanic region.

From the mouth of the geyser, a tube penetrates deeply into the earth, and fills with water, which percolates through from the surface. The water near the base of this tube is considerably heated by hot lava, but the tube is so long and narrow that convection cannot take place freely; the temperature of the water down below thus continues to rise, while that at the top of the tube is relatively cool. At the same time, the Boiling Point of the water at the base of the tube is raised by pressure, and the temperature has to reach well above 100° C. before steam is formed. When this does take place, the water above is forced upwards and flows away. The sudden reduction of the pressure at the bottom of the tube lowers the Boiling Point of the water, and the latter is now rapidly converted into steam, which forms in such volume that the entire column above it is ejected well into the air. The action of a geyser is thus intermittent. The best-known geysers are those of New Zealand, Iceland, and Yellowstone National Park, U.S.A. The famous Old Faithful geyser of Yellowstone National Park used to erupt with great precision every 66½ minutes, but is now acting with less regularity and at longer intervals.

GEYSERITE. A deposit of silica produced from a *Hot Spring* or *Geyser*. See *Sinter*.

GIBLI. The hot, dry, southerly *Sirocco* wind of Libya, North Africa.

GLACIATION. The covering of an area, or the action on that area, by an ice-sheet or by glaciers.

GLACIER.* A mass of ice which moves slowly down a valley from above the *Snowline* towards the sea under the force of gravity. It is formed owing to the pressure of the immense depth of snow, which depresses the Freezing Point; the snow in the lower layers thus melts, and then solidifies again into granular ice, or *Névé*, being later transformed by the continued pressure into clear ice.

From the snowfield, where it almost always originates, the glacier may extend far below the snowline. At its source, there is an abundant supply of ice and melting is slow, and the glacier is thus broad and deep; lower down the temperatures are higher and the supply of ice is smaller. The glacier thus decreases in size till the amount of ice melting equals the supply from above, and there the glacier ends. Usually, then, the glacier is tongue-shaped, being broadest near its source and narrowest where it finishes. Its general shape also conforms to that of the valley in which it lies. It curves with the valley, and to some extent its surface follows that of the valley floor. The centre of the glacier is usually higher than its sides, and the cross-section of its surface is thus slightly convex. The surface is often uneven, and sometimes cut by deep cracks or *Crevasses*: one cause of the unevenness of the surface is the melting of the top layer of the ice during the summer, when the water wears channels in the ice. Again, the lower layers of ice may be protected from melting by stones or other solid matter, and mounds of ice covered with debris are thus formed on the surface. See *Glacier Table*.

The rate of movement or flow of a glacier is greatest in the middle, for it is retarded by friction at the bottom and sides, and the end of the glacier is convex. Along its course the glacier collects rock material, which it deposits when the ice melts; these accumulations of fragments are called *Moraines*. See also *Piedmont Glacier*.

GLACIER BREEZE. A cold breeze of *Katabatic* type blowing down the course of a glacier, caused by the cooling of air in contact with the ice.

GLACIER MILL. See *Moulin*.

GLACIER SNOUT. The arch at the end of a *Glacier* from which flows a stream formed by the melting ice.

GLACIER TABLE.* A block of rock standing on a pedestal of ice on the surface of a glacier; the rock protects the ice beneath it from melting so quickly as the surrounding ice, and thus becomes perched on a pedestal.

GLACIOLOGY. The study of ice and the action of ice in all its forms, and therefore including *Snow*.

GLACIS. A gently inclined slope or bank.

GLADE. An open space or passage in a wood or forest, either natural or produced by the felling of trees.

GLAZED FROST. A coating of smooth ice, sometimes of considerable thickness, which forms on objects on to which rain is falling in an atmosphere having a temperature below the Freezing Point. It is formed, too, when a warm, damp wind follows an extremely cold spell, the moisture condensing and then freezing on surfaces which are still below the Freezing Point. It is comparatively rare in Great Britain, but when it does occur it may be sufficiently heavy to bring down telegraph and telephone wires.

GLEI SOIL. A type of soil which is formed under the influence of poor drainage; the *Water Table* stands at or near the surface, and a so-called glei horizon is developed, greyish in colour and characterized by reddish-brown deposits of ferric hydroxide.

GLEN. In Scotland, a long, narrow valley, with steep sides, usually having a river, a stream, or lake at the bottom. It is narrower and steeper-sided than a *Strath*, though the two terms are sometimes used indiscriminately.

GLOBIGERINA OOZE. A chalky deposit which covers a large proportion of the ocean bed, consisting chiefly of the calcareous shells of foraminifera, the globigerina being the most abundant and widely-distributed genus. The shells are minute, commonly about the size of a pin-head. Mixed with them are small amounts of the shells, etc., of other living organisms, and small traces of material derived from the land. Globigerina ooze is the most widespread type of ooze in the Atlantic and Indian Oceans, and covers much of the southern Pacific Ocean. It is most developed at depths of about 1,500 to 2,500 fathoms.

GNEISS. One of an extensive series of rocks with a banded structure, of *Metamorphic* type, varying widely in chemical composition, and usually characterized by the most conspicuous dark mineral present, e.g. hornblende gneiss. The banding, due to the separation of light and dark minerals in crude layers, is the most characteristic feature of a typical gneiss.

GNOMONIC PROJECTION. A type of *Zenithal Projection*, in which the projection is made on to the tangent plane from the centre of the sphere. Areas and shapes are so distorted that it would be altogether valueless except for the useful quality that every *Great Circle* on the sphere is represented by a straight line on the map; this enables great circle routes to be easily plotted.

GOODE'S INTERRUPTED HOMOLOSINE PROJECTION. A type of equal-area or *Homolographic Projection*, in which the oceans are interrupted but the continents are clear and well-shaped. The parallels of latitude are straight lines. Such equal-area projection of the whole earth's surface is valuable for showing the world distribution of climatic regions, economic data, etc.

GORGE. A valley which is more than usually deep and narrow, with steep walls; there is no sharp distinction between a gorge and a *Canyon*, though the latter is generally of much greater size. The sides of a small gorge are sometimes nearly vertical.

GRABEN (German). See *Rift Valley*.

GRADIENT, PRESSURE. See *Pressure Gradient*.

GRANITE. One of a number of coarse-grained *Igneous Rocks* which always contain the two minerals *Quartz* and *Feldspar*, as well as other minerals. Its structure is so coarse that the different mineral grains can be plainly seen and distinguished from one another. It is one of the hardest and most durable rocks, and also one of the most abundant rocks in the earth.

GRAPE BELT. A strip of land about 60 miles long and several miles wide extending along the southern shore of Lake Erie, U.S.A., having a long, mild autumn with unusually late frosts, and a less severe winter than areas in the same latitude but farther inland, owing to the tempering influence of the lake.

GRASSLANDS. Those regions of the world where the natural vegetation consists of grass; the rainfall is too light to permit forest growth, but is less scanty and irregular than that of the deserts, and the grasslands are thus normally situated between the forest belts and the arid regions. Two main sub-divisions are recognized, the *tropical grasslands* or *Savanna* and the *mid-latitudes*, *temperate*, or *intermediate grasslands*, of which the *Steppes*, *Prairies*, and *Pampas* are examples.

GRAVEL. A deposit of rounded stones, usually mixed with finer material such as sand or clay, and formed by the action of moving water – by a river or a lake, or by the sea. When a gravel is consolidated, it becomes a *Conglomerate*.

GREAT CIRCLE. A circle on the earth's surface whose plane passes through its centre, and therefore bisects it into two hemispheres. Two opposing meridians together form a great circle; the equator is also a great circle. The shortest distance between any two points on the earth's surface is the arc of the great circle which passes through them. See *Great Circle Route*, *Small Circle*.

GREAT CIRCLE ROUTE. A route between any two points on the earth's surface which follows the *Great Circle* between them; it is thus the shortest distance between the two points, and is often followed by ships at sea, provided that it does not take them into dangerous waters. A great circle route between two places of approximately the same latitude proceeds north of the parallel of latitude in the northern hemisphere, and south of the parallel of latitude in the southern hemisphere.

GREEN FLASH. The green light from the sun observed in a clear atmosphere at sunrise and sunset, lasting for two or three seconds. The blue and green light in the sun's rays is refracted more than the yellow and red light, and when the sun is so low that no red light from it can reach the observer, only a green light is seen; the blue light is generally absorbed or diffracted by the atmosphere.

GREEN MUD. A marine deposit which contains much fine clay and also some grains of a mineral, glauconite, a silicate of iron and potassium, from which it derives its green colour. The shells of foraminifera, whose remains are found in the green mud, are often filled with glauconite. Green mud is found particularly on the continental slope off sub-tropical coasts where few large rivers enter the sea: e.g. off the coasts of south-eastern U.S.A., California, and Portugal.

GREENSAND. A marine deposit similar to *Green Mud*, in which the glauconite predominates; it is found in the same areas as green mud.

GREENWICH MEAN TIME OR G.M.T. The *Standard Time* for the British Isles and western Europe, being the *Local Time* at Greenwich, London, on the zero meridian. For convenience, e.g. in the preparation of an airline timetable, or in the collection of simultaneous meteorological reports, G.M.T. is often used as the standard time for areas far distant from the Greenwich meridian.

GREGALE. A strong wind from the NW. or NE. direction in the south central Mediterranean area, blowing mainly in the cool season. It occurs most frequently when pressure is high to the north over central Europe and the Balkans and relatively low to the south over Libya; this type brings little or no rain and the temperature does not fall much below the normal. A similar name (e.g. grégal) has been applied to strong winds from a similar direction in other Mediterranean areas.

GRIKE or CLINT. A fissure in the surface of the rock in a limestone region, formed when the limestone dissolves in rain water containing carbon dioxide from the atmosphere; it is often formed when an open joint is widened by solution. The term is especially used of such fissures in NW. England.

GRIT. A rock, mainly of quartz, resembling a *Sandstone*.

GROTTO. A large cave produced in a limestone region by the solvent action of underground streams and percolating water; the term is sometimes loosely applied, too, to other types of cave.

GROUND FROST. In the British Isles, a temperature on the grass of approximately $-1°$ C. or below (more strictly $-\cdot9°$ C. or below); $0°$ C. ($32°$ F.) is not taken as the limiting temperature, as the tissues of growing plants are not injured until the temperature has fallen well below the Freezing Point of water.

GROUNDWATER. Water which exists in the pores and crevices of the earth's *Mantle Rock*, having entered them chiefly as rain water percolating through from the surface – as opposed to the rain water which runs off in streams. See *Run-Off*. Most of the ground water occurs within a few hundred feet of the earth's surface, for at great depths the pressure must be so great that all pores and crevices are closed. It has important chemical effects upon the rocks which it penetrates, for it may dissolve some of the rock material, it may deposit some of the mineral matter with which it has been overcharged elsewhere, or it may modify minerals by chemical combination. See *Water Table*.

GROVE. A small, shady wood; a cluster of trees set about a small open space.

GROWING SEASON. That part of the year in a specific region when the growth of the indigenous vegetation is made possible by the favourable combination of temperature and rainfall. In general, its duration decreases with distance from the equator: in the *Equatorial Forest*, for instance, the growing season is continuous throughout the year, while in the *Tundra* it lasts for only two or three months. Many cultivated plants demand growing seasons of special character; cotton, for example, requires about 200 frostless days, maize about 150 frostless days.

GROYNE or GROIN. A low wall built out into the sea, more or less perpendicular to the coast-line, to resist the travel of sand and shingle along a beach, or to minimize erosion by the sea.

GULCH. A narrow, deep ravine with steep sides formed by a torrent, chiefly applied to such a ravine in the western United States.

GULF. A large, deep bay; an extensive inlet penetrating far into the land. It may have been formed either by fracture of part of the earth's crust, or by the sea overflowing depressed land.

GULLY.* A long, narrow channel worn out by the action of water, particularly on a hillside; it is smaller than a *Ravine*, and much smaller than a *Valley*. The term is often used, too, of a channel produced in *Soil Erosion*; such a gully normally carries water only during or immediately after rain or the melting of snow.

H

HAAR. A sea fog which is driven over the eastern coastal districts of Great Britain by an easterly wind, most frequently during the summer.

HABITAT. The natural environment of a plant or animal; it is often expressed as one of the main natural regions which are recognized in geography, or one of their sub-divisions, e.g. the equatorial forest, the tundra.

HABOOB. A type of duststorm experienced in north and north-east

Sudan, being most frequent near to Khartoum. It is commonest in the afternoon and evening, and is rare in the morning; it occurs chiefly between May and September, but is possible at any time of the year. The haboob is usually accompanied by a sudden increase in wind strength and a change of direction, a marked fall in temperature, and an extremely poor visibility caused by the raising of the dust; it is often followed by heavy rain and sometimes thunderstorms.

HACHURES. Short lines of shading drawn on a map to represent differences in the slope of the ground. If the slope is steep, the lines or hachures are thick and close together; if the slope is gentle, they are thinner and farther apart. The darkest portions of a hachured map, then, represent crags and very steep slopes, while the lighter shades depict the gentler slopes; the white parts show plateaux, level mountain tops, hillside shelves, or level valley bottoms. Such a map, hung at a distance from the eye, thus gives a fairly good impression of the configuration of the land. The hachures are drawn from the highest point of a mountain to the foot, and should be perpendicular to the *Contour* lines. Unlike the contour lines, they give no indication of the actual height of land above sea level.

HADE. The angle between a *Fault Plane* and the vertical.

Haffs along the southern shores of the Baltic Sea.

HAFF (German). A *Lagoon*, formed at the mouth of a river owing to the presence of a sandy bar or *Spit*, and usually so shallow as to be accessible only to small vessels: applied especially to the Baltic coast of the former East Prussia (now East Germany and Poland), where the spit is known as a *Nehrung*.

HAIL. The hard pellets of ice which fall from *Cumulonimbus* clouds, and

GEO—6

are often associated with thunderstorms. They are of various shapes and may reach considerable size; individual hailstones have been known to weigh over 2 lb. They are caused by the rapid ascent of moist air; the water drops freeze, and the size of the pellets increases as more water vapour freezes on to their surface. When the pellets are heavy enough to overcome the resistance of the ascending air currents, they fall; in their downward journey, they may still grow by assuming fresh layers of ice from the supercooled water drops in the moist air. A severe hailstorm is capable of causing immense damage to growing crops. *Soft hail* consists of pellets of ice crystals which resemble snow.

HALO. A ring of light which surrounds the moon or sun when the sky is covered by a thin white veil of *Cirrostratus* cloud. When the solar halo is formed, it is rarely seen with the naked eye owing to the brightness of the sun, but is visible through darkened glass. The commonest type is that in which the angular distance between moon and halo is 22°. It is formed by refraction of light through the ice crystals of the cloud. The halo is usually white, but if the ring is very sharply defined it is coloured, having a reddish tinge on the inner edge.

HALOPHYTE. A plant which grows naturally in a *Salt-Marsh* or other saline environment; it is often adapted, like the *Xerophytes*, to conserve water

HAMLET. A small village or group of houses in a rural area.

HAMMADA or HAMADA. The rocky uplands of a desert, particularly in the Sahara, which have been swept clear of sand and dust by the wind.

HANGING VALLEY. The valley of a tributary which enters a main river valley from a considerable height above the bed of the latter, and so forms rapids or waterfalls down the slope. Hanging valleys are particularly common in glaciated regions; when a glacier has eroded the main river valley, the opening of a tributary into the latter is left at some height above the new, lower valley. They are not always due to glaciers, however; they are formed whenever the main river valley has been eroded more quickly than the tributary valleys.

HARBOUR. A stretch of water on the coast which affords shelter to sea-going vessels; it may have been formed naturally, or artificially, or by artificial improvement of a natural feature. See *Port*.

HARDPAN. A hardened or cemented layer of soil, impervious to drainage, lying below the surface.

HARMATTAN. A strong easterly or north-easterly wind experienced in West Africa. Blowing direct from the Sahara, it is hot, very dry, and dusty – so hot and dry that it sometimes splits the trunks of trees. When it penetrates to the Guinea coast, it provides a welcome

relief from the moist heat, and is beneficial to health; it is a relatively cool wind to that area, in fact, for by reason of its extreme
dryness it promotes evaporation and therefore cooling. See *Doctor,
The*. On the other hand, it carries with it abundant dust from the
desert, and thick haze is sometimes formed and impedes river navigation; it may cause severe damage to crops; far inland, it is so dry
and dusty as to be often injurious to health. Its average southern
limit in mid-winter, when the equatorial belt of low pressure is
along or just south of the Guinea coast, is about 5° N.; in midsummer it is about 18° N. It has thus a much greater duration in
the north of the area than in the south.

HAZE. A mass of minute solid particles of dust, smoke, etc., which obscures the atmosphere near to the earth's surface so as to reduce the
visibility below 2 kilometres, but not below 1 kilometre. See *Mist, Fog*.

HEADLAND. A steep crag or cliff jutting out into the sea.

HEAD OF NAVIGATION. The farthest point up a river to be reached
by vessels for the purposes of trade.

HEATH. An extensive tract of open, uncultivated ground, more or
less flat and covered with small shrubs such as heather or ling.

HEAT WAVE. A term commonly used to signify an unbroken spell of
abnormally hot weather. In the British Isles, for instance, a period
of several successive days when the temperature exceeded 27° C.
would constitute a heat wave.

HELIUM. One of the rare gases contained in the atmosphere, all of
which together constitute only about .01 per cent of dry air; the
proportion of helium is about one part in 200,000.

HELM WIND. A strong, cold, north-easterly wind which blows down
from the summits of the hills in northern England, especially down
the western slopes of the Cross Fell Range in Cumberland and
Westmorland. When it blows, a heavy, helm-shaped cloud, the
'Helm', forms over the Cross Fell Range, and a few miles away and
parallel to the 'Helm' appears a thin roll of cloud known as the
'Helm Bar'; the Helm Wind blows as far as the 'Helm Bar'. The
Helm Wind usually occurs when the general direction of the wind
is easterly or north-easterly, and is most common in late winter
and spring.

HEMISPHERE. One half of the earth's surface, formed when a plane
through its centre bisects the earth. The earth is usually divided
into northern hemisphere and southern hemisphere, the former
being that half of the earth's surface which is north of the equator,
the latter that half which is south of the equator. See *Land Hemisphere, Water Hemisphere*.

HIGH. A region of high atmospheric pressure. See *Anticyclone*.

HILL. A small portion of the earth's surface elevated above its sur-

roundings, of lower altitude than a *Mountain*. In general, an eminence is not considered a mountain unless its elevation, from foot to
summit, is well over 1,000 ft, but the distinction is arbitrary.

HILLOCK. A mound or small *Hill*.

HINTERLAND. The land which lies behind a seaport or seaboard and
supplies the bulk of the exports, and in which are distributed the
bulk of the imports of that seaport or seaboard. The seaport to
which an inland region is the hinterland is sometimes determined
by the outline of the land. The West Riding of Yorkshire, England,
for example, is part of the hinterland of Liverpool for trade with
America, but for trade across the North Sea is part of the hinterland
of Hull, Goole, and Grimsby. Sometimes the hinterlands of different seaports overlap, even in relation to the same sea; this may be
due to variations in the facilities for shipping at the seaports and in
facilities for communication with the hinterland; Hull, for instance, is
capable of taking large sea-going ships, while Goole, only accessible
to smaller vessels, has the advantage of nearness to their hinterland.
The extent and importance of a hinterland may be much increased
by improvements in its seaport or in inland transport facilities.

HOAR FROST. The deposit of ice crystals formed on the surface of
objects near to the ground when water vapour in the atmosphere
solidifies without first passing through the liquid state; it is formed
instead of *Dew* when the *Dew Point* is below the Freezing Point. It
often consists partly of frozen dew and partly of ice formed directly
from water vapour.

HOGBACK or HOG'S BACK. A long, narrow ridge in which both *Dip*
slope and *Scarp* slope are steep, owing to the steep inclination, or
dip, of the beds, and thus differing from a *Cuesta*.

HOMOLOGRAPHIC or EQUAL-AREA PROJECTION. A type of *Map
Projection* in which the area bounded by any two adjoining parallels
and two adjoining meridians is equal to any other area similarly
enclosed; in other words, the ratio between any area on the map
and the corresponding area on the globe is constant.

HOMOSEISMAL LINE or COSEISMIC LINE. A line drawn through
places which are affected by an earthquake at the same time. Such
lines are usually elliptical, the mid-point of the ellipse being the
Epicentre.

HOODOO. See *Earth Pillar*.

HOOK. A *Spit* which is curved at one end; the curvature may be
caused by the action of the waves in rolling material to the sheltered
side of the spit. (See Fig. p. 85.)

HORIZON, RATIONAL or TRUE. The *Great Circle* which is parallel to
the sensible or visible horizon of an observer. See *Horizon, Sensible
or Visible*.

HORIZON, SENSIBLE or **VISIBLE.** The more or less circular line bounding an observer's view of the earth (or sea), where earth (or sea) and sky appear to meet. On land the line is broken by surface obstructions; at sea the line is a perfect circle, the observer being at the centre. The distance of the horizon is determined by the height of the observer. Allowing for refraction, and ignoring obstructions, the actual distance of the sensible horizon from an observer placed at an elevation of 100 ft is about 14 miles; this distance varies as the square root of the height of the observer.

Cape Cod, Massachusetts, showing the Hook at its tip.

HORIZON, SOIL. A layer of soil which lies more or less parallel to the surface and has fairly distinctive soil properties. See *Profile, Soil.*

HORSE LATITUDES. The sub-tropical belts of high atmospheric pressure over the oceans, situated in both hemispheres between the *Trade Winds* and the *Westerlies.* They move north and south with the sun. They are regions of calms and light variable winds, of comparatively dry air, and quiet, stable weather conditions. The name may have originated from the practice, in the days of sailing ships, of throwing overboard horses which were being transported to America or the West Indies when the ship's passage was unduly prolonged.

HORST. An elevated block of rock between parallel *Faults* which has reached its position either through uplift between the faults, or through the sinking of the beds outside the faults.

HORTICULTURE. The cultivation of flowers, fruit, or vegetables, usually on small plots of land, and thus a form of *Intensive Cultivation*.

HOT SPRING or THERMAL SPRING. A stream of hot water issuing from the ground, often after being heated by *Magma*, and therefore commonly occurring in a volcanic region when eruptions have ceased. Hot springs are not confined to volcanic regions, however, as water sinking far enough into the earth may become heated and rise to form springs. Mineral substances which have been held in solution are often deposited around the hot springs. See *Fumarole, Geyser, Travertine*.

HUERTA. Land in Spain which, owing to prolonged high temperatures and irrigation, produces two or more crops per year; it is situated mainly in the south-east, in the provinces of Valencia, Murcia, and Granada, and yields abundant oranges, pomegranates, figs, and almonds, occasionally even sugar-cane and cotton. The efficient cultivation of the huerta contrasts with the poverty and neglect of much of the interior plateau of Spain. See *Vega*.

HUM. In a *Karst Region*, a small mass of limestone standing above the surrounding country because of greater resistance to erosion. The term is also applied to a remnant of limestone which is resting on non-calcareous rocks and is being gradually worn away by solution.

HUMIDITY. The state of the atmosphere with respect to the water vapour it contains. See *Absolute Humidity, Relative Humidity*.

HUMUS. The decomposed and partly decomposed organic matter, of animal and vegetable origin, in the soil. In uncultivated land the humus is derived from the natural decay of previous generations of plants; in ploughed and cultivated land the humus is supplied as some kind of manure. The humus of ordinary soil is black, and is thus responsible for making the soil darker than the subsoil. It plays an important but very complicated part in maintaining the fertility of the soil. The amount of humus in different soils varies considerably; some, like the peat soils, consist largely of slightly decomposed organic matter which has not yet become humus.

HURRICANE. (1) A wind of Force 12 on the *Beaufort Scale* of wind strengths: i.e. a wind having a mean velocity of over 75 miles per hour.

(2) The *Tropical Cyclone* or revolving storm of the West Indies and Gulf of Mexico. It usually originates east of the islands, occasionally as far east as the Cape Verde Islands, and takes a westward course, sometimes causing extensive destruction on one island after another; it then generally recurves to the north-east. Besides the

West Indies, the Gulf coasts of the United States and the eastern side of Central America as far south as the Mosquito Coast of Nicaragua are affected by hurricanes; Costa Rica, Panama, and the northern coast of South America, however, lie outside the hurricane region. The months of greatest frequency are September and October. The name is also given to the tropical cyclone experienced off the coast of Queensland, Australia.

HYDRO-ELECTRIC POWER. Energy obtained from natural or artificial waterfalls and utilized for the generation of electricity in dynamos worked by turbines.

HYDROGRAPHY. The science that treats of the waters of the earth's surface, particularly with reference to their physical features position, volume, etc., and the preparation of charts of seas, lakes, rivers, contours of the sea bed, shallows, deeps, currents, etc.

HYDROLOGY. The science that treats of water, especially in relation to its occurrence in streams, lakes, wells, etc., and as snow, and including its discovery, uses, control, and conservation.

HYDROSPHERE. The comparatively shallow layer of water which covers over two-thirds of the earth's surface, and forms the oceans, seas, and lakes; through it projects the *Lithosphere*, forming the continents.

HYDROPHYTE. See *Hygrophyte*.

HYETOGRAPH. A self-recording instrument, very similar to a *Rain Gauge*, which registers the rainfall. The collected rainfall is usually made to raise a float, to which is attached a pen writing on a chart that is wrapped round a clock-driven drum.

HYGROGRAM. The continuous record of the *Relative Humidity* of the atmosphere, as measured by the *Hygrograph*, usually to cover a period of a week.

HYGROGRAPH. A self-recording *Hygrometer*, in which the *Relative Humidity* of the air is continuously traced on a chart fixed to a rotating drum, actuated by clockwork. The ordinary hair hygrograph, which is in common use at meteorological stations, depends on the expansion and contraction in length of human hair with variations in Relative Humidity, these changes being then transferred to the rotating chart, or *Hygrogram*, by a pen. It is not an accurate instrument, but is useful for recording any considerable or sudden changes in the amount of moisture in the atmosphere, and the time at which they occur.

HYGROMETER. An instrument designed to measure the *Relative Humidity* of the atmosphere. It usually consists of two thermometers, a *dry bulb thermometer*, measuring the actual temperature of the atmosphere, and a *wet bulb thermometer*; the bulb of the latter is kept permanently moistened by a piece of muslin wrapped round it and an attached length of wick dipping into a reservoir of water. Evaporation from the wet bulb causes the temperature of the wet bulb thermometer to be lower than that of the dry bulb thermometer, and the difference between the two temperatures – sometimes known as the wet bulb 'depression' – affords a measure of the Relative Humidity of the air. More complicated forms of hygrometer have been devised, but are not in general meteorological use. See *Hygrograph, Psychrometer, Wet Bulb Temperature*.

HYGROPHYTE or HYDROPHYTE. A plant which thrives only when a considerable amount of moisture is available, and therefore lives either in water or in a very humid region. The stem is usually long and comparatively fragile, and contains little woody fibre; the leaves are large and thin, and the roots shallow. The banana tree provides an example.

HYGROSCOPE. An instrument for showing the changes in the humidity or dampness of the air, usually by means of the changes in appearance or dimensions of a substance. In a crude wall instrument like the 'weather-house', containing the figures of a man and a woman, the state of the air is indicated by the twisting and untwisting of a piece of catgut in response to changes of humidity.

HYPABYSSAL ROCKS. Those *Igneous Rocks* which have risen towards the earth's surface, but have failed to reach it, solidifying in fissures as *Dykes, Sills, etc.*

HYPSOGRAPHIC CURVE. A diagrammatic curve in which heights are plotted against areas, often being used to represent the average form of the earth's surface: the various heights of the land and the depths of the ocean are represented by vertical distances above and below a horizontal line, which signifies mean sea level, and the areas which the heights and depths cover on the earth's surface are represented by lengths along this horizontal line. Heights and depths are much exaggerated. The salient facts illustrated by the curve are: first, predominance of two levels representing the general bed of the oceans and the general flat surface of the continents, and second, the steepness of the *Continental Slope* joining the two predominant levels.

HYPSOMETER. An instrument which measures altitude. Usually it is an instrument in which the boiling point of water is accurately measured; as this temperature varies with the atmospheric pressure, the latter can be calculated from it, and from this the altitude can be obtained. In order to obtain the height to within 10 ft, the temperature of the boiling point of water must be measured to within one hundredth of a degree.

HYTHERGRAPH. A graphical representation of the differentiation between various types of climate. Mean monthly temperatures are plotted as ordinates against the mean monthly rainfalls as abscissae, and a closed, twelve-sided polygon, the hythergraph, is obtained. This reveals the type of climate at a glance. See *Climograph*.

I

ICE AGE. A geological period in which ice sheets and glaciers covered large areas of the continents, reaching the sea in places and lowering the temperature of the oceans. The latest Ice Age began early in the Quaternary Period (See *Cainozoic Era*), when the ice covered much of Europe and North America; the present *Ice Sheets* of Greenland and Antarctica are relics of this Ice Age.

ICEBERG. A mass of land ice which has been broken off or 'calved' from the end of a glacier or from an ice barrier, and is afloat in the sea. When a glacier enters the sea, the ice is buoyed up by the water, and a portion of the glacier is easily broken off and floats away. A glacier berg is irregular in shape; a berg from an ice barrier is rectangular in shape, flat-topped or 'tabular', often very large, and is characteristic of the Antarctic.

The main sources of icebergs are the great ice sheets which cover Greenland and Antarctica. The Greenland icebergs, which break off from glaciers, are carried south by the Labrador Current towards the Grand Banks, off Newfoundland. They often project 200 or 300 ft above the surface, and are sometimes several hundred yards long. The icebergs from the Antarctic ice barrier are sometimes over 40 miles long and project 300 to 400 ft above the surface. The number of icebergs varies considerably from year to year; when they are particularly numerous, they may be encountered far beyond their usual limits.

Only about one-ninth of the mass of an iceberg is visible above the water, so that the depth of a berg below the surface is far greater than its height above the surface. An iceberg gradually decreases in size owing to melting, erosion, and *Calving*. The melting is caused both by the sun and by the entry of the iceberg into warmer waters. Erosion is caused by the swell, the waves, and rain. Calving often disturbs the balance of the iceberg and causes it to roll over. An iceberg often carries with it boulders and other rock material which came from the glacier; when melting takes place in the water, this debris sinks to the sea bed.

ICEBLINK. A white glare on the horizon caused by the reflection of light from a mass of ice which is too far away to be visible.

ICE-CAP. See *Ice Sheet*.

ICE-CAP CLIMATE. The type of climate to be found over the polar ice-caps or *Ice Sheets* of Greenland and Antarctica, which may be regarded as cold deserts of snow and ice; it is the least known of climatic types, and data are sparse. The mean monthly temperature throughout the year is below 0° C. The mean annual temperature of the interior of the Greenland ice-cap is estimated at −32° C., and that of the Antarctic interior is probably lower. The small precipitation consists almost entirely of dry, granular snow. In Antarctica, especially towards the margins of the ice-cap, severe blizzards occur, with wind speeds well over 50 miles per hour.

ICE-FIELD. A uniform, unbroken ice-floe of great extent; a continuous sheet of ice formed when lumps of *Pancake Ice*, having increased in size, finally join up.

ICE-FLOE. A mass of floating ice, detached from the main polar ice, and of large or small extent. A *light floe* is one in which the ice is only two or three feet thick, a *heavy floe* one in which the ice is thicker.

ICE-FOOT. A mass of ice projecting into the sea on an Arctic or Antarctic shore. It sometimes becomes very thick, with its upper edge several feet above sea level. The first stage in its formation takes place during the autumn, when snow accumulates along the shore; water thrown up by the waves freezes on to this snow, and forms a mass of ice. This is later augmented by lumps of sea-ice, likewise forced on land by tides and waves. Fragments of rock often collect on the ice-foot, and protect the ice from melting so that parts of it remain till the following autumn.

ICE SHEET or ICE-CAP. A vast mass of ice and snow which covers large land areas in the polar regions; its surface is almost flat. The ice sheets of Greenland and Antarctica are the only large ones now in existence. The former covers the whole of Greenland except for a narrow fringe; only along this fringe does any uncovered rock appear. The Antarctic ice sheet, unlike that of Greenland, extends beyond the land surface which it covers, and projects out to sea in some parts. The thickness of the Greenland ice-cap and of the Antarctic ice-cap has been estimated to be several thousand feet in places. Sometimes the Greenland ice-cap is described as an 'Arctic Sahara', or a 'cold desert'; it is more of a desert than the Sahara, however, for it is even more deficient in plant and animal life. See *Ice-Cap Climate*.

IGNEOUS ROCKS. Rocks which have solidified from a molten *Magma*, and form one of the three main types of rocks which comprise the earth's crust. They may have solidified after reaching the surface (*Lavas* or volcanic rocks), or in channels connecting the molten reservoirs with the exterior (*Hypabyssal Rocks*), or well below the surface under pressure (*Plutonic* or *Abyssal Rocks*). In many of them the various minerals have crystallized separately, and the rock is a mass of interlocking crystals. They do not usually occur in distinct beds or strata, and they are not fossiliferous. There are many kinds of igneous rocks, owing to the varying conditions under which the original magma solidified. See *Metamorphic Rocks*, *Sedimentary Rocks*.

IMPERMEABLE ROCKS. Rocks which, being non-porous or practically so, do not allow water, e.g. rain water, to soak into them. Granite provides an example; this rock may be pervious, however, owing to joints and fissures. Some geographers assume the term to be synonymous with *Impervious Rocks*.

IMPERVIOUS ROCKS. Rocks which do not allow water, e.g. rain water, to pass through them freely; they may be porous, like clay, or practically non-porous, like unfissured granite. Some geographers use the term *Impermeable Rocks*.

IMPORTS. Goods which are brought into a country in trade, being obtained from another country which has a surplus; the usual reason is that the first country finds them impossible, or difficult, or uneconomic, to produce at home. Sometimes climate prevents or restricts the domestic production of goods; tropical fruits, for instance, cannot be grown in the intermediate zone. Sometimes land and labour can be more profitably devoted to the production of other goods, and the commodities are imported; wheat and meat, for instance, are leading imports into Great Britain. In general, imports may be regarded as goods obtained in exchange for *Exports*.

INCISED MEANDER. See *Entrenched Meander*.

INDIAN SUMMER. A spell of calm, mild weather sometimes experienced in late autumn in Great Britain and in the northern United States. The term is supposed to have originated in North America towards the end of the eighteenth century; it is attributed to the fact that the area where the weather conditions were first observed was still inhabited by American Indians.

INLET. A small opening into the coastline or into the bank of a lake or river.

INLIER. A mass of old stratified rocks surrounded by newer strata. See *Outlier*.

INSELBERG (German). An 'island mountain', an isolated steep hill, a type of *Relict Mountain* occurring in a hot dry region.

INSOLATION. The radiant energy received from the sun by the earth and other planets. It varies considerably over different parts of the earth's surface. The amount of insolation reaching any place during any one day depends on the *Solar Constant*, the area of the surface and its inclination to the sun's rays, the transparency of the atmosphere, and the position of the earth in its orbit. This amount varies throughout the day with the changes in the angle of incidence of the sun's rays; and the length of time that the sun remains above the horizon is a determining factor. At the *Equinoxes*, that time is twelve hours per day everywhere on the earth's surface, but towards the summer *Solstices* the longer days of the higher latitudes more than compensate for the greater obliqueness of the sun's rays. Thus the total insolation per day is not at its maximum at the latitude where the sun is highest at noon, but at a rather higher latitude where the daily duration of sunlight is greater. For the year as a whole, however, insolation is greatest at the equator; it decreases at first slowly, then more rapidly, then slowly again towards the poles. Insolation shows the least variation throughout the year at the equator, and varies very considerably at the poles.

INSULAR CLIMATE. The type of climate experienced on islands and in coastal regions where the sea rather than one of the great land masses is the dominating influence. In contrast to the *Continental Climate*, the insular climate is thus characterized by a relatively small range of temperatures, both diurnally and from summer to winter. See *Maritime Climate*.

INTENSIVE CULTIVATION. A system of farming by which the cultivator expends much labour and possibly capital on a relatively small area. Land may be kept continually under cultivation, without being allowed to lie fallow at any time. Intensive cultivation is practised in countries such as Denmark and England, where land is relatively valuable, and high crop yields are sought and obtained. See *Extensive Cultivation*.

INTERFLUVE. The ridge between two adjacent river valleys.

INTERIOR DRAINAGE. See *Internal Drainage*.

INTERLOCKING SPURS. In a V-shaped *Valley* down which a river with a winding course is flowing: the portions of the valley walls which project from both sides to the concave bends of the river and so obscure the view upstream.

INTERNAL or INTERIOR DRAINAGE. Drainage in which the waters have no outlet, and so do not reach the sea. An example is the Great Basin of Nevada and western Utah, U.S.A., where the streams flow into basins and create *Salt Lakes*, from which the water continually evaporates or sinks into the ground.

INTERNATIONAL DATE LINE. The line approximating to the meridian 180° W. (or E.), where the date changes by exactly one day as it is crossed; a traveller crossing the date line westwards adds a day, while one crossing it in an eastward direction subtracts a day. This is because on a journey eastward from the Greenwich meridian, local time advances till 180° E. is reached, when it is 12 hours ahead of G.M.T.; similarly, moving westwards, local time is retarded till

180° W. is reached, when it is 12 hours behind G.M.T. Thus there is approximately 24 hours' difference in time between two points placed just each side of 180° W. (or E.) longitude – the International Date Line. Some deviations of the line from the 180° meridian have been agreed upon in order to avoid confusion of dates in island groups and land areas that are cut by the meridian.

INTERTROPICAL FRONT or INTERTROPICAL CONVERGENCE ZONE. The *Front* or boundary where the north-east and south-east *Trade Winds* meet, being thus associated with the *Doldrums*. As the term 'front' for this boundary is not strictly correct, the term 'intertropical convergence zone' is preferable.

INTRUSIVE ROCKS. Rocks which solidified from *Magma* before reaching the earth's surface; they are thus *Igneous Rocks*. See *Extrusive Rocks*.

INVERSION, TEMPERATURE. An increase of temperature with height above the earth's surface, being the reverse of the normal situation, in which the temperature falls with height. It is commonly experienced in hollows and valleys, especially in winter on calm, clear nights, when radiation has caused considerable cooling and the cold air has sunk down into them; a pool of cold air thus lies there, while on the mountain slopes above the air is warmer. When the sky has been clear and the wind light for some time, it is also usual in fairly level areas in temperate latitudes for a temperature inversion to develop above the surface at night. In winter, during anticyclonic weather, the inversion may reach a considerable height, and may persist for several days; the dense winter fogs of some cities (e.g. London) generally originate from this type of inversion.

INVISIBLE EXPORTS. Items representing services rendered to a foreign country which take the place of normal exports, but do not involve the actual transfer of goods. Examples are the trade conducted for foreign countries by national shipping, expenditure in the country by foreign tourists, the interest on national capital invested in foreign countries.

IRRIGATION. The artificial distribution of water on the land in order to facilitate the cultivation of crops where otherwise, owing to a deficiency of rainfall, agriculture would be impossible. In regions bordering rivers which regularly overflow their banks, such as the Nile, irrigation simply consists in the provision of canals and sluices to control the flow of water over the land surface. In some areas the ancient method of baling water from a well by hand is still practised. It is usual, however, for a dam or barrage to be built across a river, creating an artificial lake; canals from this lake, then branch canals and distributaries, lead the water under control over the cultivated ground. This method is widely utilized in the United States, India, and elsewhere. Irrigation water can be much more profitably used than an equal amount of rain water, for it can be stored until needed, instead of much of it being allowed to sink into the ground to a great depth and be lost to agriculture. There is a smaller loss, too, from evaporation. In Persia evaporation is almost entirely prevented by distributing the water through underground channels known as ganats.

Although costly, irrigation has many advantages. It ensures a regular supply of water when most needed, and thus increases the productivity of the land; it helps to fertilize the land by distributing sediments; it often permits more valuable crops to be grown than would be possible by utilizing only rainfall; it often permits cultivation for a longer period, and enables several crops to be grown. On much of the irrigated land of the world, the density of population is therefore high.

ISALLOBAR. A line drawn on a *Weather Chart* through places at which the same change of pressure or *Pressure Tendency* has taken place during some period of time; it is obtained by plotting the change in atmospheric pressure which has taken place between two sets of observations. Isallobars are drawn to indicate regions of rising and of falling atmospheric pressure.

ISANOMALOUS LINE. A line on a map joining places having equal departures from the normal in some meteorological element. A map of isanomalous lines joining places of equal *Temperature Anomaly*, for example, shows at a glance the regions which are abnormally warm for their latitude, having a positive temperature anomaly, and those which are abnormally cold for their latitude, having a negative temperature anomaly. As the effects of latitude and altitude are eliminated, such a map gives a simple picture showing the influence of the land and sea, ocean currents, and prevailing winds.

ISLAND. A piece of land surrounded by water, in an ocean, sea, lake, or river. It may be formed in one of several different ways, e.g. by *Diastrophism*, or movement of the earth's crust: the sea bed may have risen or the sea subsided till parts of the bed have emerged, or part of a coastal plain may have been submerged by the rise of the sea or the sinking of the land. Alternatively, a volcano may have built up its cone till the top projects above the sea; a coastal region may have been eroded by the sea or by glaciers; sediment may have been deposited along the sea shore or in a river to form a low, sandy island; the island may have been built up by coral polyps. Again, two or more of these different processes may have combined to form an island.

Islands in the sea are often classified as *continental* or *oceanic*. A continental island is one which was formed by separation from the mainland, an oceanic island one which was formed in the ocean, independent of the mainland. An island which was volcanic in origin is called a volcanic island, and one which was built up by coral polyps a *Coral Island*.

ISLET. A small island.

ISOBAR. A line on a map joining places having equal atmospheric pressure. To render the pressure readings comparable with one another, they are usually corrected by reducing to mean sea level. In climatological studies, isobars may be drawn to show the average pressure distribution in an area or throughout the world during a certain period; they may also be drawn on *Weather Charts*, to show the pressure distribution at a certain time on a certain day. In the latter case they are almost invariably drawn at intervals of 2 millibars; the isobars drawn are those of even number, 1,000, 1,002, 1,004 millibars, etc., the position of these readings being found by interpolation.

ISOBATH. A line on a map joining points on the sea bed which have equal depth. Such lines thus show the relief of the sea bed, just as *Contours* show the relief of the land by joining places of equal altitude.

ISOCLINAL FOLDING. The folding of rock strata in which the *Folds* are so tightly packed together that the limbs of each fold dip in the same direction.

ISOGONIC LINE. A line on a map joining places having an equal magnetic declination. See *Declination, Magnetic*.

ISOHALINE. A line on a map joining points in the oceans which have equal *Salinity*.

ISOHEL. A line on a map joining places having equal duration of *Sunshine*.

ISOHYET. A line on a map joining places having equal depth of *Rainfall* over a certain period.

ISONEPH. A line on a map joining places having equal average *Cloudiness* over a certain period.

ISOPLETH. A line on a map drawn through places having the same value of a certain element. Such lines show the geographical distribution of the elements. Familiar examples of isopleths are *Isobars, Isohyets, Isotherms*.

ISOSEISMAL LINE. A line on a map joining places which have suffered an equal intensity of shock from an earthquake. Such lines usually form closed curves round the *Seismic Focus*. They are often very irregular, for earthquake damage depends to some extent on the type of foundations on which buildings rest, as well as on their distance from the origin. If a number of isoseismal lines are drawn, however, it is usually possible to determine the approximate position of the *Epicentre*.

ISOSTASY. The state of equilibrium that is said to exist between the highlands and the lowlands of the earth, due to the fact that the former are made of lighter rock materials than the latter. According to the theory of isostasy, known as the isostatic theory, the *Continental Platforms* are made of lighter material and are, as it were, floating at a higher level than the material of the ocean floor.

ISOSTATIC THEORY. See *Isostasy*.

ISOTHERM. A line on a map joining places having the same temperature at a particular instant, or having the same average temperature over a certain period; the temperature is normally reduced to mean sea level in order to eliminate differences due to altitude. The isotherms most commonly used in geography are those joining places having the same mean monthly temperature, especially for January and July, which show the average temperature conditions in midwinter and midsummer. The isotherms of mean daily maximum and mean daily minimum temperatures are also extremely useful.

ISTHMUS. A narrow strip of land joining two large land areas, e.g. two continental land masses, or joining a peninsula to the mainland; examples are the Isthmus of Panama and the Isthmus of Suez.

J

JEBEL or JABAL (Arabic). A mountain or range of mountains.

JOINT. A crack in a mass of rock which has been formed along a plane of weakness or *joint plane*; unlike a *Fault*, little or no movement has taken place between the blocks. Where the rock is exposed to the weather, joints have much influence on the shape of cliffs and crags. In sedimentary rocks, usually one set of joints is parallel to the *Dip* of the rock strata and another set is parallel to the *Strike*. The joints of igneous rocks are variable in form. *Basalt*, for instance, often breaks into hexagonal columns, as on the Giant's Causeway, Ireland, while there are often also transverse joints across the columns.

JUNGLE. Wild, uncultivated land with a dense undergrowth; the term is often applied to the typical forests of *Monsoon* lands.

K

KAINOZOIC ERA. See *Cainozoic Era*.

KAME. A mound of gravel and sand which was formed by the deposition of the sediment from a stream as it ran from beneath a glacier. Kames are thus often found on the *Outwash Plain* of a glacier.

KAOLIN. See *China Clay*.

. Typical BAD LANDS scenery, showing hillsides lined with GULLIES,
n South Dakota, U.S.A.

. View of Crater Lake, Oregon, the best-known CALDERA in North
merica. The small island is a volcanic cone formed after the original
olcanic summit had sunk.

3. Part of the Grand Canyon of the Colorado River, Arizona, U.S.A., th
largest CANYON in the world; the river is just discernible at its foot.

A CREVASSE on the Forno Glacier, Switzerland. Photo: E. Meer-ämper, Sils-Engadin.

5. The Pyla Dune, in the LANDES, S.W. France. Photo: J. Belin.

6. The famous Old Faithful GEYSER, Yellowstone National Park, Wyoming, U.S.A., in eruption.

7. GLACIER TABLES on the Aletsch Glacier, Switzerland. Photo: Otto Furter, Davos-Platz.

8. The LEVEE built at Greenville, Mississippi, U.S.A., to protect factories and farms from the flood waters of the Mississippi River. The high level of the river, on the right, will be noted.

9. An aerial view of the hole near Winslow, Arizona, U.S.A., caused by a
METEORITE; it is nearly a mile long and about 600 feet deep.

10. The MIDNIGHT SUN, seen from within the Arctic Circle in
northern Norway.

11. The Fiescher GLACIER, Switzerland, showing medial and lateral *moraines*. Photo: Otto Furter, Davos-Platz.

12. Sand DUNES Near Colomb Bechar, Algeria, illustrating their en-croachment on an OASIS of palm trees. Photo: Ofalac, Algiers.

13. A scene in North Dakota, U.S.A., on the level, almost treeless PRAIRIES of North America; in the background a DUSTSTORM may be seen approaching.

14. A remarkable SCARP formation, produced by EROSION and bounded by numerous TALUS slopes: the Great Organ and Temple, Capitol Reef National Monument, Utah, U.S.A.

15. Devil's Tower, Wyoming, U.S.A., a VOLCANIC NECK, consisting of solidified LAVA, which is 600 feet from base to summit; it was left behind when the remainder of the cone was worn away by WEATHERING.

16. SERACS on the Grindelwald GLACIER in Switzerland.

7. The STALAGMITE known as the Speaker's Mace in Cox's Cave, Cheddar, Somerset, England, and other stalagmites and STALACTITES; the formations are also seen reflected in a pool on the floor of the cave.

18. A scene in Java, showing Terrace Cultivation of rice fields.

19. A small section of the great CONIFEROUS FOREST region of Canada, showing the TIMBER LINE in the background, and, at a still higher level, the SNOWLINE. Canadian Pacific Photograph.

20. The CRATER of the
Raoeng VOLCANO, Java.

21. An aerial view of Niagara Falls, the best-known WATERFALL in the world. Canadian Pacific Photograph.

22. An ARID region of Arizona, U.S.A., where the rainfall is so deficient that the only plants are XEROPHYTES such as these cacti.

KARABURAN. The hot north-easterly wind experienced in the Tarim basin of Sinkiang, when the interior of the Asiatic land mass is strongly heated; it sets in during the early spring in Central Asia, and continues by day till the end of summer. It is strong, often blowing with gale force, and sweeps up clouds of dust from the desert, darkening the air and causing acute discomfort. The coarse sand is not carried beyond the deserts, but the lighter dust particles are conveyed long distances, cause a characteristic haze, and on settling on the ground form *Loess*. The sand blown along by the karaburan is one of the principal causes of the changes in the courses of rivers through the desert.

KARREN. See *Lapiés*.

KARRENFELD (German). In *Limestone* country, an area in which the rocks exhibit ridges and deep flutings, being known as karren or *Lapiés*.

KARST REGION or KARSTLAND. A limestone region in which most or all of the drainage is by underground channels, the surface being dry and barren: named after the Karst limestone district of the Dinaric Alps, near the Adriatic coast of Yugoslavia. The calcium carbonate in the limestone is carried away in solution, and only the insoluble material is left to form a covering to the rocks; the soil is thus usually thin, and the surface bare, except in the valleys, where a greater depth of soil may accumulate. In the Karst region of the Dinaric Alps, where this type of country is particularly well developed, the rivers are hidden except where the roof of a cave has collapsed, forming a *Sink Hole*. The rain water of such a region tends to disappear at once into the underground channels, and it is the solution of the limestone by this water, before and after it sinks, which gives rise to the uneven topography of the region – the typical Karst topography. Many short gullies and valleys end suddenly where the water is discharged into caves or subterranean channels.

KATABATIC WIND. A local wind caused (often at night) by the flow of air, cooled by radiation, down mountain slopes and valleys; it is also caused by the flow of cold air down the slopes of ice-caps, such as those of Antarctica and Greenland. With the rapid loss of heat by radiation, the mountain or ice-cap becomes cold, and the air above it is chilled and moves downwards under the action of gravity. The direction of flow is controlled almost entirely by orographic features. See *Mountain Wind, Anabatic Wind*.

KETTLE HOLE or KETTLE. A hollow or depression in an *Outwash Plain*, probably formed because an ice block has been covered by gravel borne by streams emerging from the glacier, and had subsequently melted, allowing the debris to settle.

KHAMSIN. The hot, dry, southerly wind experienced in Egypt, corresponding to the *Sirocco* of North Africa. According to the Arabs, it blows during a period of fifty days, from April to June, the name khamsin being the Arabic word for fifty. It blows ahead of the depressions which move eastwards along the Mediterranean or across North Africa, and often carries with it a considerable quantity of dust from the interior. In the Middle East the name is also rather loosely applied to any hot, dry wind blowing off the desert.

KNICK POINT. A break in the profile of a river (See *Profile, River*), which may be due to uplift of the land, causing the river to be *Rejuvenated*.

KNOLL. A small, rounded hill or mound.

KNOT. A unit of speed equal to one *Nautical Mile* per hour. In the days of sailing ships, the knot was a length of knotted string, attached to the log-line; the number of knots, arranged at regular intervals along the line, which ran off the reel in a certain time indicated the ship's speed in nautical miles per hour.

KOEMBANG. A *Föhn* type of wind experienced in Java, which sometimes damages sensitive crops such as tobacco.

KOPJE or **KOPPIE** (Afrikaans). In South Africa, a hill.

L

LACCOLITH or **LACCOLITE.** A large mound of *Igneous Rock* beneath the earth's surface, sometimes attaining the size of a mountain; it is caused by the intrusion of *Magma* from below into the crust of the lithosphere without reaching the surface, the surface strata being

Section through a Laccolith.

arched up over the intrusion. It is formed when the magma is viscous, and instead of spreading widely, heaps up at a particular level. Well-formed laccoliths have a flat bottom and a dome-shaped upper surface, but they are often very irregular. See *Sill*.

LACUSTRINE. Relating to a *Lake*.

LAGOON. A shallow stretch of water which is partly or completely
separated from the sea by a narrow strip of land. In the case of a
Coral Reef, it is the channel of sea water between the reef and the
mainland, or, when an *Atoll* has been formed, the sheet of water
enclosed by the latter; a fringing reef, close to the shore, has a relatively narrow and shallow lagoon, while a barrier reef, far from the
shore, has a much wider and deeper lagoon. A lagoon may also be
formed by a spit of land, composed of mud, sand, or shingle,
closing or almost closing the entrance to a bay – e.g. off the Nile
delta. Again, it may be formed when the sea throws up a bank of
shingle at high water mark, and encloses a sheet of water between
the bank and the cliff; the lagoon then sometimes disappears at low
tide. A lagoon may be formed, too, when an arm of the sea is enclosed by sandhills. See *Haff*.

The lakes of Finland, resulting from glaciation.

LAKE. An extensive sheet of water enclosed by land, occupying a
hollow in the earth's surface. The name is sometimes loosely
applied, too, to the widened part of a river, or to a sheet of water
lying along a coast, even when it is connected with the sea; there
are many gradations, in fact, between bays and lagoons which are
almost enclosed and coastal lakes.

Usually the amount of water entering a lake exceeds that lost by
evaporation, and there is an outflowing stream; the water of the

lake is thus fresh. In a region of low rainfall and great evaporation, however, the lake has no outlet, and it forms an inland drainage area. All the salts brought down in solution by the rivers accumulate in such a lake, which thus acquires a very high *Salinity*; the Dead Sea and the Great Salt Lake, Utah, U.S.A., are examples of this kind of lake. See *Salt Lake*. When rainfall is seasonal, the level and area of a lake may fluctuate considerably; Lake Chad, in Africa, for instance, decreases greatly in size during the dry season. A lake may also dry up entirely during a drought or in the dry season, leaving only a salt-covered mud flat; Lake Eyre and other lakes of central Australia disappear in this way. A lake remains permanent provided that the amount of water it receives, as rain and as water draining into its hollow, equals the amount lost.

LAKE DWELLING. A prehistoric human shelter, raised on a platform supported by piles above the surface of a lake, characteristic of the Neolithic Age in Switzerland. Similar dwellings are still used in various parts of the world, e.g. in New Guinea, Borneo, the estuary of the Amazon.

LAKE RAMPART. A wall-like mass of stones round the margin of a lake or gulf, formed in a climate where the winter range of temperature is great. When a thick layer of ice lies on a mass of cold water, the temperature of the under-surface of the ice is kept nearly constant owing to the presence of the water, while the upper surface is exposed to great variation owing to the changes in air temperature. The upper surface expands and contracts, and thus sometimes cracks and sometimes buckles up, according as it is contracting or expanding. If the shore on which it is resting is low, the expanding ice is pushed over its margin, and, besides modifying this margin considerably, often pushes over it loose stones from the shallow water. As it retreats again, the ice leaves these stones behind, and they ultimately form the wall-like ramparts.

LAMBERT'S AZIMUTHAL EQUAL-AREA or EQUIVALENT AZIMUTHAL PROJECTION. An equal-area or homolographic hemispherical *Map Projection*. The relative sizes of the areas are correctly shown, but except at the equator there is a north-south extension and an east-west compression of the right and left margins as compared with the centre. Besides being a homolographic projection, it is also azimuthal – all points have their true compass directions from the centre of the map. The projection is commonly used in atlases to show the hemispheres.

LAND BREEZE. A diurnal wind blowing from the land out to sea, caused by the differential cooling of land and sea: the cooling of the air over the land by radiation during the night causes the air to descend and flow seawards. It is confined to coastal regions and lake sides, and, though most characteristic of tropical lands, may also be observed in the temperate zone during the summer. It is most developed when the general *Pressure Gradient* is slight and the sky clear; it may then set in about midnight or a few hours later. The land breeze is much influenced by topography, and varies considerably in different parts of a coast. It often alternates with the *Sea Breeze*, but is usually less developed than the latter.

LANDES.* Low-lying, sandy plains bordered by sand *Dunes*, especially applied to the district of this name in SW. France.

LAND HEMISPHERE. That half of the globe, mainly situated north of the equator, in which the great majority of the land surface, about six-sevenths, is contained. Such a high proportion of the land surface may be included by constructing the hemisphere with its centre at London, or Berlin, or some neighbouring point in northern Europe.

LAND-ICE. Ice formed inland from fresh water, in contrast to ice formed from sea water.

LANDSLIDE or LANDSLIP. The downward movement of a large mass of earth or rocks from a mountain or cliff. It is often caused by rain water soaking into the soil and earthy material on a steep slope; their weight is much increased, and they become more mobile. It may also be caused by an earthquake, or on the sea coast by the undermining action of the sea.

LAPIÉS (French). The surfaces of limestone rocks in a *Karst Region* which are grooved and fluted owing to solution by rain water containing carbon dioxide, the grooves and flutings being the channels where the latter runs down the rock faces. The corresponding German term is karren.

LAPILLI. Small rock fragments ejected from a volcano; they are smaller than *Volcanic Bombs*, but larger than *Volcanic Ash*, being generally considered to vary in size from peas to walnuts. They are also known as cinders.

LAPSE RATE. The rate of change of temperature with height in the atmosphere, being usually expressed in degrees Fahrenheit per thousand feet or degrees Centigrade per hundred metres. It is taken as positive when the temperature decreases with height, as it normally does, and negative when the temperature increases with height, as in a temperature inversion. See *Inversion, Temperature*. The average lapse rate in the atmosphere is about ·6° C. per hundred metres.

LATERITE. A red, ferruginous rock characteristic of damp tropical regions. It is relatively soft when first quarried, but hardens on exposure to the atmosphere. It may be rich enough in iron to be used as a source of that metal, and it may be rich in alumina. It occupies considerable areas of peninsular India, Malaya, and Indonesia, and equatorial Africa.

LATITUDE. The angular distance of a point on the earth's surface north or south of the equator, as measured from the centre of the earth. *Parallels* of latitude are lines drawn round the earth parallel to the equator, and may thus be described as approximate circles with the two poles as centres, the circles becoming smaller with increasing proximity to the poles. As one-quarter of the complete circumference of the earth is traversed in moving from the equator to the pole, one right-angle or 90° being covered, the parallels of latitude are marked off in ninety divisions or degrees from the equator to each of the poles; the equator represents 0° latitude, while the N. pole is 90° N. latitude, and the S. pole 90° S. latitude. Each degree is sub-divided into 60 minutes, and each minute into 60 seconds. Degrees of latitude are equivalent to approximately the same linear distance, 69 miles, everywhere on the earth's surface; near the poles, owing to the flattening of the earth, they are slightly longer.

LAVA.* Molten rock or *Magma* which has flowed from the interior of the earth on to its surface, through the crater of a *Volcano* or through fissures in its side. On the surface the molten material solidifies more quickly than in the interior of the earth. Some lavas contain much silica and are said to be 'acid', others have little and are termed 'basic' lavas. The acid lavas have a high melting point, are very viscous, flow slowly, and do not travel far; the basic lavas have a lower melting point, are very fluid, move rapidly, and may flow for several miles before solidifying. Acid lavas thus lose their gases the less readily, and cause the volcanoes to erupt explosively, the whole mass sometimes bursting into fragments or even dust; basic lavas give comparatively quiet eruptions, with a widespread flow of lava. Most volcanoes, however, are of an intermediate kind. The surface of solidified lava is usually very rough; often in both acid and basic lavas the escaping gases cause the upper layers of the stream to be vesicular. See *Scoria*.

LEACHING. The process by which material such as organic and mineral salts is washed out of a layer of soil into a lower layer by percolating rain water.

LEAP YEAR. A *Year* of 366 days, occurring every fourth year, and devised in order to allow for the fact that the duration of the earth's revolution round the sun is about 365¼ days; by the addition of a complete day every fourth year, or leap year, the correction necessary to cause the seasons to fall in the same months of every year is approximately made. See *Calendar*.

LEEWARD. The side or direction which is sheltered from the wind, e.g. by a range of mountains.

LENTICULAR CLOUD. A lens-shaped type of cloud, most often seen over hills or mountains. It has sharp edges, and the numerous cloudlets of which it is formed are often seen streaming through the cloud. It is frequently associated with a *Föhn* wind.

LESTE. A hot, dry, southerly to easterly wind experienced in Madeira ahead of an advancing depression; it is often dust-laden, and resembles the *Leveche*.

LEVANTER. The easterly wind sometimes experienced in the Straits of Gibraltar and in southern Spain, named after the direction from which it blows. It causes dangerous eddies on the lee side of the Rock of Gibraltar. With a moderate wind strength, a *Banner Cloud* extends from the summit of the Rock a mile or more to leeward; when the wind has gale force, the cloud lifts and disappears. The wind is also known as the Solano.

LEVECHE. The hot, dry, southerly wind experienced in SE. Spain, corresponding to the *Sirocco* of North Africa; it blows ahead of an advancing depression, and often carries with it much dust and sand.

LEVEE.* The natural bank of a river formed during flooding by the deposition of sediment. When the flood subsides, the sediment remains, and the levee is thus the highest portion of the *Flood-Plain* of a river. The height and length of a levee may be sufficient to divert the course of a tributary. A levee continues to be raised by flooding, and as the river also goes on depositing material in its own channel, and raises the bed, both levee and river bed may finally lie above the adjoining country. This is the case in the middle and lower reaches of the Mississippi. The levees along the lower reaches of the River Po, in northern Italy, too, are so high that the surface of the river is above the roofs of neighbouring villages. Occasionally a flood may be heavy enough to break down levees and inundate large areas of land. A city or town is sometimes built on a levee, New Orleans being an example; there the natural levee was only a few feet above the river level on one side and the swamp on the other side. Artificial levees are constructed so as to supplement the work of the natural levees in keeping the water of such rivers as the Mississippi within its proper channel.

LEVEL OF SATURATION. See *Water Table*.

LEY FARMING. A method of farming in which *Arable Land* is sown with grass and is then maintained as pasture for a number of years.

LIANA OR LIANE. A tropical climbing plant common in the *Equatorial Forest*. The stem of the liana, which is usually thick and woody, climbs and twists round the trunk and branches of a tree till the summit is reached, and sometimes kills it. Even when the tree has rotted away, however, the liana is still able to hold itself in place by its lateral branches, which extend for long distances through the boughs of other trees.

LIGNITE OR BROWN COAL. A coal of woody appearance, usually considered to represent the second stage in the transformation of vegetable matter through *Peat* into coal. Lignites and brown coals are sometimes classified as separate types.

LIMESTONE. A rock consisting essentially of calcium carbonate. There are many types of limestone, all of which have certain common characteristics owing to their similar chemical composition, and are distinguished by name from one another, e.g. oolitic limestone, carboniferous limestone. Most of them are of organic origin, and contain the hard parts of various organisms such as the shells of molluscs and the skeletons of corals. Limestone, unlike most rocks, is readily soluble in ground water which contains carbon dioxide. See *Cave, Karst Region, Stalactite*.

LINE SQUALL. A line of *Squalls* of considerable length, sometimes several hundred miles, occurring simultaneously, its passage usually being marked by a line of low, dark cloud and a sudden rise in wind speed, often with other violent changes in weather conditions. Its position marks the replacement of a warm air current by a much colder air mass, and it is accompanied by the typical *Cold Front* of a depression; the ascent of the warm air over the cold air produces the cloud, together with heavy rain or hail; and a veer of wind, a rise in pressure, and a fall in temperature occur at the passage of the front.

LINGUA FRANCA. A common language which is spoken or under-stood over a wide area by people of various races, although not their native tongue. English is the most important lingua franca, being spoken or understood over more than half the world. Other examples are Mandarin Chinese, Malay, and Swahili. See *Pidgin English*.

LITHOSPHERE. The solid crust which envelops the inner *Barysphere* of the earth. It consists of the thin, loose layer known as the soil and the mass of hard rock, several miles in thickness, upon which the soil lies. It projects through the *Hydrosphere* to form the continents.

LITTORAL. The seashore, the strip of land along a sea coast, or, more strictly, the land lying between high and low tide levels; pertaining to the seashore. The littoral region of the *Ocean* comprises the shallow waters adjacent to the sea coast; this region has the richest vegetation, and so supports the most abundant animal life. In the littoral region of the temperate zone, sometimes known as the *Continental Shelf*, the world's most important fishing grounds are situated.

LLANOS. The *Savanna* of the Orinoco basin and the Guiana highlands, north of the equatorial forests of the Amazon basin in South America. In Spanish the word means 'plains'.

LOAD, RIVER. The solid matter carried along by a river, including (1) material in solution in the water, (2) material suspended in the water, mainly mud, silt, and sand, and (3) the larger, heavier material carried along the bed of the river. The material in solution is sometimes excluded from the load. The maximum or full load of a river depends on its velocity and its volume, and on the size of the particles constituting the load; when the limit to the possible load has been reached, any further addition involves the dropping of an equivalent portion of the original load.

LOAM. A rich soil consisting principally of a mixture of sand and clay, together with *Silt* and *Humus*; it has the good qualities of both sand and clay, but not their bad qualities.

LOCAL TIME. The time at any point on the earth's surface calculated by the position of the sun; noon local time is thus the moment when the sun reaches its highest position in the sky. In view of the rotation of the earth on its axis, places lying on the same *Meridian* will thus have the same noon and the same local time. Of two places on different meridians, however, the more easterly will have a local time preceding the other's: i.e. at any given instant its clocks will be 'fast on' or 'in front of' those of the more westerly place. As the earth rotates through 360° once in 24 hours, local time changes by one hour for every 15° longitude, or 4 minutes for every 1° longitude.

LOCH. In Scotland, a lake, fiord, or arm of the sea. The most numerous of the Scottish lochs are the small *Tarns*, which lie in hollows formed by glaciation or by the damming effect of moraines left behind after the glaciers have retreated. More important are the larger *Glen* lochs, which lie in hollows in the glens, and the long, narrow sea lochs, or *Fiords*, which are especially common along the western Highland coast.

LODE. A composite mineral *Vein*, consisting of a number of parallel fissures filled with mineral matter which contains *Ores*.

LOESS. A deposit of fine silt or dust which is generally held to have been transported to its present situation by the wind. One characteristic is its ability to maintain vertical walls in the banks of streams, and another is the number of small vertical tubes running through it, probably due to the remains of roots. It covers wide areas round the margins of the arid interior of Asia, having been transported there by winds which have swept along the fine dust from the dry plateaux. In northern China, where it has enormous extent and thickness, the loess is a fine *Loam*, rich in lime and yellowish in colour. Its great advantage is its extreme porosity, for rain sinks quickly into it and its surface is normally dry, but when irrigated it is very fertile. Through the loess, streams have cut deep valleys, often with vertical walls. There are also extensive loess deposits in central Europe and the central United States, and, mixed with a high proportion of *Humus*, it forms the fertile *Black Earth* of the U.S.S.R.

LONGITUDE. The angular distance, measured along the equator, between the *Meridian* through a given point and a standard or prime meridian; since 1884, the meridian through the former Royal Astronomical Observatory at Greenwich, in London, has been accepted by international agreement as the prime meridian. Longitude is thus measured in degrees west or east of Greenwich, from 0° to 180°, the Greenwich meridian representing longitude 0°; the meridian 180° W. coincides with meridian 180° E. The complete angular distance round the earth round any parallel of *Latitude* is 360°, so if 360 meridians are drawn from pole to pole at equal intervals, they will lie 1° of longitude apart. As the parallels of latitude became shorter with increasing nearness to the pole, it follows that a degree of longitude also decreases towards the poles. It is longest at the equator, where it approximates to a degree of latitude, about 69 miles; at 30° N. or S. the length of a degree of longitude is about 60 miles, at 60° N. or S. 34.6 miles, at 80° N. or S. 12 miles, and at the poles it is zero. The longitude of a place may be found from the difference between *Local Time* and *Greenwich Mean Time* or G.M.T., as shown by *Chronometers*.

LONGITUDINAL VALLEY. A valley which runs parallel to conspicuous mountain chains. See *Transverse Valley*.

LONGSHORE DRIFT. The movement of shingle and sand along a sea shore, due to the waves advancing obliquely up the beach.

LOUGH. In Ireland, a lake, fiord, *Ria*, or arm of the sea.

LOW. A *Depression*, or a region where the atmospheric pressure is lower than that of its surroundings. On a *Weather Chart*, the centre of a depression is usually indicated by the word 'low'.

LOWLAND. The low-lying land of a region, in contrast with the mountainous areas.

LOXODROME. See *Rhumb-Line*.

LUNAR ECLIPSE. See *Eclipse, Lunar*.

LUNAR MONTH. The interval of time in which the *Moon* makes one complete revolution round the earth, from new moon to new moon: approximately 29.5 days. It is often popularly taken to be a period of 28 days.

M

MACCHIA. See *Maquis.*

MACKEREL SKY. A sky covered with *Cirrocumulus* or *Altocumulus* clouds, i.e. with small, high, rounded and detached cloud masses with blue sky in the gaps, somewhat resembling the markings on the side of a mackerel. It is usually seen in summer, during dry warm weather.

MAELSTROM. A large *Whirlpool.* The term has been applied specifically to the Moskenstrom, a strong current between Mosken and Moskenaes, two of the Lofoten Islands, off the west coast of Norway.

MAESTRO. A north-westerly wind experienced in the central Mediterranean area, most strongly on the western side of a depression: especially so-called in the Adriatic Sea, the Ionian Sea, and round the coasts of Corsica and Sardinia.

MAGMA. The molten material which exists below the solid rock of the earth's crust, and sometimes reveals itself on its emission from a *Volcano.* It does not always reach the surface in a molten state through a volcano, however, but may cool and solidify as it forces its way upwards. See *Dyke, Sill, Laccolith, Batholith.* Magmas vary considerably in chemical composition; those which contain more than about 50 per cent silica are commonly called acid, those which contain less than 50 per cent silica are called basic.

MAGNETIC DECLINATION. See *Declination, Magnetic.*

MAGNETIC POLE. See *Magnetism, Terrestrial.*

MAGNETIC STORM. A large, irregular variation or disturbance of the earth's magnetic field, of world-wide effect, mainly or completely masking the normal diurnal variations. See *Magnetism, Terrestrial.* It may last for a few hours or a few days, and may occur at any time of the year. It is associated with *Sunspots,* and occurs almost simultaneously with them; the *Aurora* is usually observed in low latitudes during a magnetic storm, and at these times there is considerable interference with radio and telegraphic communications.

MAGNETISM, TERRESTRIAL. The magnetic properties displayed by the earth, causing it to act like a huge bipolar magnet; the science of the study of these properties. At any point on the earth's surface, a magnetic compass, allowed to swing freely in a horizontal plane, will come to rest pointing approximately in the direction of the North and South Poles; the actual points which the compass needle indicates, known as the North and South Magnetic Poles, lie respectively NNW. of Boothia Peninsula, North America, and in Victoria Land, Antarctica. Although a compass is used to find true north, it therefore points somewhat east or west of true north at most places. See *Declination, Magnetic.*

A number of theories have been evolved to explain the cause of terrestrial magnetism, but none adequately explains all the observed facts. Besides the irregular variations of the earth's magnetic field, known as *Magnetic Storms,* there is a diurnal variation, in which the compass swings through an angle of several minutes, and a secular variation of several degrees, which occupies a period of many centuries.

MAIZE RAINS. One of the two heaviest falls of rain experienced during the year in East Africa, occurring between February and May. See *Millet Rains.*

MALLEE SCRUB. The *Scrub,* consisting largely of low eucalyptus bushes, which is characteristic of the dry sub-tropical regions of south-east and south-west Australia.

MANDATED TERRITORY. A territory which was transferred after World War I from either the German or the Turkish Empire to the League of Nations, and was thereafter governed, under Mandate of the League, by one of its member states. Territories such as Iraq, Syria, and Palestine were to be administered by the Mandatory Powers only until they were able to govern themselves. Some territories, such as the former German colonies in Africa, were to be administered by the Mandatory Power but were to retain their entity, while others, such as South-West Africa and former German colonies in Australasia, were to be governed by the Mandatory Power as part of its territory. After World War II all Mandated Territories except South-West Africa became *Trust Territories.*

MANGO SHOWERS. See *Blossom Showers.*

MANGROVE SWAMP. A swampy area occupied chiefly by mangroves, occurring mainly in the low coastal lands of tropical regions, most extensively near river mouths; from the branches of the trees descend long arching roots, which anchor the trees and form an almost impenetrable tangle at ground level.

MANTLE ROCK or REGOLITH. The layer of loose fragments, the surface part of which is called soil, which covers most of the earth's land area, and varies in thickness from place to place.

MAP. The representation on a flat surface of all or part of the earth's surface, to show physical, political, or other features, each point on the diagram corresponding to a geographical position according to a definite scale or projection. See *Map Scale, Map Projection.* The most accurate maps of Great Britain are the Ordnance Survey Maps, prepared by surveyors from actual measurements.

MAP PROJECTION. A method by which the curved surface of the earth is represented on a flat surface map, so that each point of the latter corresponds to one point only of the former. It involves the construction on the plane surface of a graticule formed by two intersecting systems of lines, corresponding to the parallels of latitude and the meridians of longitude on the earth. Some projections aim at showing directions correctly, other areas or shapes, and for general purposes the most useful projections are those which take all of these factors into consideration. It is impossible, however, to construct a map of a considerable portion of the earth's surface on a flat surface without some distortion of shapes, relative areas, or directions; the only true representation of the earth is a globe. In atlases, *Cylindrical Projections* and *Conical Projections* are commonly used.

MAP SCALE. The ratio between the distance between two points on a map and the actual distance it represents. It may be expressed by the two lengths, e.g. one inch to one mile; this would be for a large-scale map, for the lengths on the map representing certain actual distances are relatively large. Such a large-scale map shows surface features in great detail, and could be constructed only in a country which had been well surveyed; British Ordnance Survey Maps, for instance, are on this scale. (1 inch to 1 mile corresponds to a scale of 1 : 63,360). A map scale may be expressed, too, as an arithmetic ratio, e.g. 1 : 2,000,000; this would be for a small-scale map, such as is used in atlases, any length on the map being small in relation to the actual distance it represents. (1 : 2,000,000 corresponds to a scale of 1 inch to about 32 miles.)

MAQUIS (French) or MACCHIA (Italian). The low *Scrub* of parts of the Mediterranean region, consisting of small trees or shrubs which are able to withstand drought; among them are the wild olive and the myrtle. The maquis is of little use to man, for it is difficult to clear for cultivation. It corresponds to the *Chaparral* of California.

MARBLE. A coarse-grained, crystalline *Metamorphic Rock* derived from *Limestone*.

MARES' TAILS. Wispy, tufted *Cirrus* clouds.

MARITIME CLIMATE. A climate which is mainly influenced by the proximity of the sea, giving a comparatively cool summer and a comparatively warm winter. This is due to the different thermal capacities of land and water; the sea is warmed and cooled less readily than the land, and its influence tends to create an equable climate. See *Insular Climate, Continental Climate*.

MARKET GARDENING. The *Intensive Cultivation* of vegetable crops for market. See *Truck Farming*.

MARL. A mixture of clay and calcium carbonate – though the term is rather loosely applied to a wide variety of rocks and soils. Some of the marls are marine deposits, while others are of fresh-water origin.

MARSH. A tract of soft, wet land, usually low-lying and partly or completely under water; the extreme dampness is due to the impermeable nature of the soil, such as clay, and the poor drainage. See *Salt-Marsh, Swamp*.

MASSIF (French). A mountainous mass which breaks up into peaks towards the summit, and has relatively uniform characteristics. Probably the best-known example is the Massif Central of France.

MEANDER. A curve in the course of a river which continually swings from side to side in wide loops, as it progresses across flat country: the name being derived from the river Meander of Asia Minor, which in its lower reaches has a particularly twisting course. The meander is continually being accentuated by the river itself. On the concave side of a curve the bank is worn away by the current, while on the convex side solid material is deposited. Ultimately the meander may form almost a complete circle, when the river will cut across the narrow strip of land, and this part of its course will become almost straight. See *Entrenched Meander, Ox-Bow Lake*.

MEAN SEA LEVEL or M.S.L. The average level of the sea, as calculated from a large number of observations taken at equal intervals of time. It is the standard level from which all heights are calculated. On the Ordnance Survey Maps of the British Isles, the Ordnance Survey Datum, above which the heights of all places are given, is taken as Mean Sea Level at Newlyn, Cornwall.

MEAN SPHERE LEVEL. The mean level of the *Lithosphere*. It is represented by an imaginary spherical surface in such a position that if all the land projecting above it were sliced off, this land would just fill all the depressions below it.

MEDITERRANEAN CLIMATE. The type of climate experienced by the lands bordering the Mediterranean Sea and also by other regions, in both hemispheres, situated in a similar geographical position: on the western sides of the continents, on the tropical margins of the intermediate or temperate region, and in approximately the latitude of the sub-tropical high-pressure systems – between 30° and 40° N. and S. latitude. Owing to the seasonal swing of the world pressure belts, it is dominated by the *Westerlies* in winter and the *Trade Winds* in summer; the moderate rainfall is thus received almost completely during the winter, the summer being dry. The vegetation is specially adapted to withstand the summer drought. Other characteristics are the warmth of the summer, the mildness of the winter, and the ample sunshine. It is a climate which has acquired a high reputation for the abundance of fruit and flowers which it produces, and in it many of the world's great holiday resorts, such as those of the Riviera, have been established. Typical fruits are the grape, the olive, and the citrus fruits – orange, lemon, and grape-fruit. Considerable areas are covered by small trees and shrubs (see *Maquis*). Besides the lands around the Mediterranean Sea, other areas enjoying this type of climate are central California, central Chile, the southern tip of South Africa and parts of SW. and southern Australia.

MELTEMI. The Turkish name for the *Etesian Winds*.

MENHIR. A tall, upright stone of Neolithic origin, probably set up for religious purposes, e.g., to mark a burial place.

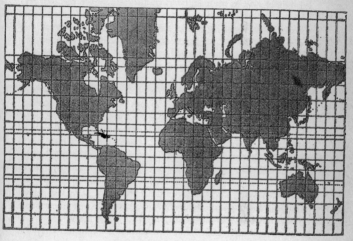

Mercator's Projection.

MERCATOR'S PROJECTION. A type of *Cylindrical Projection*, used for maps of the world, and first published in 1569. All the parallels of latitude have the same length as the equator, whereas on the globe they decrease in length towards the poles; hence there is an east-west stretching on this projection everywhere except at the equator, and this stretching increases with distance from the equator. On the globe, the parallel of latitude at 60° is one half the length of the equator, so at that latitude the stretching is two-fold; to compensate for this, a two-fold north-south stretching is also made. Thus a small area, say 1° square, at latitude 60° is on the map stretched to twice its length and twice its breadth, or four times its area compared with an area of 1° square at the equator. As the stretching, both east-west and north-south, becomes so much greater towards the poles, maps on this projection are rarely taken beyond 80°, for at this latitude stretching is about six-fold in each direction. The effect may be best observed by comparing Greenland and South America; although the former has actually less than one-twelfth the area of the latter, on Mercator's projection it is depicted as slightly the larger of the two.

There is so great a distortion of distances, areas, and shapes of land masses and oceans on this projection, in fact, that it is un-suitable for general use. One of its advantages is that the 'equal stretching' causes compass bearings to be truly represented, and makes it useful for marine charts; the compass course between any two points is the straight line joining them. This property is due to the facts that (1) the meridians are parallel straight lines, (2) the projection is *Orthomorphic*. Mercator's projection is also valuable for certain climatological maps, as, for instance, where the true directions of winds in various regions are to be represented.

MERE. (1) A small lake (as in Cheshire, England), pond, estuary, or arm of sea. In the English Fens, the term is applied to marshland. (2) A boundary or landmark.

MERIDIAN. A line of *Longitude*, or half of one of the *Great Circles* which pass through the poles and cut the equator at right-angles; it is not the complete great circle, for there is a difference of 180° longi-tude and twelve hours of time between the two halves of a great circle passing through the poles. The Greenwich meridian is the meridian passing through Greenwich, London, and is taken as the zero or prime meridian from which other meridians are measured. At the equator the distance between any two meridians, or 1° longitude, is about 69 miles; this distance diminishes with the distance from the equator, and at 50° latitude is rather less than 45 miles.

MESA (Spanish). A flat, table-like mass, which falls away steeply on all sides; the word in Spanish means 'table'. The harder top layers of rock have resisted denudation, and, being practically horizontal, have maintained a uniform surface parallel to the stratification. Mesas may be formed, as for instance in the Colorado basin of the United States, when a river cuts a deep *Canyon* through a region, and later the valley floors widen, and the plateau becomes broken up into large, individual, flat-topped masses. A mesa is related structurally to a *Cuesta*. In time a mesa becomes reduced by dis-section and erosion to a *Butte*.

MESETA. The extensive plateau of interior Spain, which covers about three-quarters of the country. It is a much eroded and broken block of the earth's crust, and is crossed by certain higher moun-tain ridges.

MESOPHYTE. A plant which requires an average amount of moisture. Most of the common trees, shrubs, and herbs living in intermediate climates of moderate rainfall and temperature are mesophytes.

MESOZOIC ERA. The Middle Life Era, the third of the five major subdivisions of the geological scale of time: that period of the earth's history between the *Palaeozoic Era* and the *Cainozoic Era*. Throughout the era there was a great deal of terrestrial disturbance, with the formation of mountains and much volcanic activity. It was characterized by the number, variety, and enormous size of its reptiles, both on land and in the sea, and is often known as the Age of Reptiles; in the middle of the era, flowering plants, deciduous trees, and the first birds appeared. It is sometimes known as the Secondary Era.

MESTIZO. The offspring of a European, usually Spanish or Portuguese, or a *Creole*, and an American Indian, or, in the case of Brazil, a Negro.

METAMORPHIC AUREOLE. The rock immediately surrounding a mass of *Intrusive Rock*, such as a *Boss*, which has undergone metamorphism, or change, by contact with it. See *Metamorphic Rocks*.

METAMORPHIC ROCKS. Rocks which were originally *Igneous* or *Sedimentary Rocks*, but have been changed in character and appearance; they form one of the three main types of rocks which comprise the earth's crust. The change in their original state may have been due to the heat, which sometimes causes the minerals to recrystallize, or pressure, which alters the rock structure, or possibly to the action of water, which dissolves some rock material and deposits other material, changing the composition of the rock. Sometimes the metamorphic rocks differ so greatly from the igneous or sedimentary rocks from which they were derived that their origin is practically impossible to determine. Most common igneous and sedimentary rocks have a metamorphic equivalent. *Granite*, for instance, may have its principal crystals separated and arranged in crude layers, to become a *Gneiss*; *Limestone* may be recrystallized to marble; *Shale* may be transformed by great pressure into *Slate*.

METEOR or SHOOTING STAR. A fragment of solid matter which enters the upper atmosphere from outer space, and becomes visible through incandescence caused by the resistance of the air to its passage. It travels up to 50 miles per second, and usually becomes luminous when it is between 50 and 100 miles above the earth's surface, disappearing at a height of about 50 miles. Little is known about the origin of meteors. Most of them are very small, and are so much disintegrated during their passage through the atmosphere that they fall to the surface as dust; occasionally, however, a larger body, known as a *Meteorite*, falls to the ground.

METEORITE.* A large mass of rock or mineral similar to a *Meteor*, but so large that it is not disintegrated to dust in the atmosphere, but reaches the earth's surface. Some meteorites are entirely metallic, consisting chiefly of an iron-nickel alloy; others contain silicates as well as metallic materials, others are almost wholly silicates. None of them has been found to contain any chemical element not previously found on the earth. One of the best-known meteorites is one which fell near Winslow, Arizona, U.S.A., and formed a basin nearly a mile in diameter and about 600 ft deep.

METEOROGRAPH. A self-recording instrument which is attached to a hydrogen-filled balloon, and thus, on the release of the latter, obtains automatic readings of certain meteorological elements, usually pressure, temperature, and humidity, in the upper atmosphere. See *Ballon Sonde*.

METEOROLOGY. The science which investigates the weather, in particular the physical processes which occur in the atmosphere, and the connected processes of the lithosphere and hydrosphere. The subject therefore includes the study of atmospheric pressure and temperature, the winds, precipitation, cloudiness, sunshine, etc.

MIDNIGHT SUN.* A phenomenon of high latitudes observed at and around midsummer, when the sun does not sink below the horizon throughout the 24 hours, and therefore may be seen at midnight; it is due to the inclination of the earth's axis to the plane of the *Ecliptic*, and the consequent tilting of each hemisphere towards the sun during its summer. At North Cape, Norway, just north of 73° N. latitude, for instance, at least part of the sun's disc is continuously visible, day and night, from May 12 to July 29.

MILE, STATUTE. A unit of linear measurement used in Great Britain and other English-speaking countries, including the U.S.A., equal to 1,760 yards; the name was derived from the Roman measure of 1,000 paces, about 1,618 yards. See *Geographical Mile, Nautical Mile*.

MILLET RAINS. One of the two maximum falls of rain experienced in East Africa, from October to December. See *Maize Rains*.

MILLIBAR. A unit of pressure equal to one-thousandth of a *Bar*. It is the unit of atmospheric pressure on the C.G.S. system, and the one usually employed by meteorologists in drawing weather charts, the *Isobars* being drawn at intervals of 2 millibars. A pressure of 1,000 millibars is equivalent to approximately 29.53 inches or 750.1 millimetres of mercury.

MINERAL. A natural inorganic substance which possesses a definite chemical composition (normally a rock does not), and definite physical and chemical properties. There are about 2,000 known minerals, with many different crystal forms, but within a single mineral species little variation occurs. Of this large number of minerals, one to two hundred are abundant, while less than a dozen are common in most rocks.

MINERALOGY. The science of the study of *Minerals*.

MINERAL SPRING. A *Spring* which contains a noticeable quantity of mineral matter in solution – except calcium carbonate or calcium sulphate; a *saline spring*, for instance, contains certain salts in solution (e.g. common salt), a *sulphur spring* contains sulphuretted hydrogen, a *chalybeate spring* contains iron compounds. A mineral spring often has medicinal value, e.g. the *Hot Springs* of Karlovy Vary, in Bohemia, Czechoslovakia.

MINUTE. (1) A unit of time equal to one-sixtieth of an hour.

(2) A unit of measurement of *Latitude* or *Longitude* equal to one-sixtieth of a degree.

MIRAGE. An optical illusion in which images of distant objects are seen, caused by the refraction of light through layers of air of different density. In a hot desert, for instance, an observer may have the illusion of a distant sheet of water, shimmering in the sunlight: the air near to the ground, owing to excessive heating, becomes less dense than the layers above; rays of light from the sky approaching the observer at a slight angle, are refracted towards him, and appear to come from the surface of a sheet of water. Other objects may appear to be floating on the water, their lower parts being invisible. A similar mirage may sometimes be seen over a smooth road surface on a hot day even in temperate regions. Another type of mirage is that in which double images of distant objects can be seen. Sometimes a mirage is formed, too, when light rays are bent downwards from a warm layer of air resting on a cooler, denser layer; this is frequently seen in polar regions, and causes ships or icebergs to appear as if inverted and suspended in the sky, though nothing is visible on the sea.

MISFIT RIVER. A river that appears to be too small to have carved out the valley through which it flows. This may be due to the fact that it has lost its headwaters by *River Capture*.

MIST. A mass of water drops present in the lower layers of the atmosphere, caused by the condensation of water vapour in the air, and giving a moderate reduction of visibility. According to meteorological practice, such a phenomenon remains a mist as long as the visibility exceeds 1,000 metres or 1,100 yards, but becomes a *Fog* when the visibility falls below this limit.

MISTRAL. A cold, northerly or north-westerly wind experienced on the shores of the north-west Mediterranean, especially around the Rhone delta. It is most prevalent during the winter, when pressure is relatively high over the European continent and low over the western Mediterranean; it reaches great force in the rear of a deep depression situated over the Gulf of Lions which is moving away eastwards or south-eastwards. The air sweeps southwards from the Central Plateau of France, is funnelled through the Rhone valley, and thus reaches the delta as an extremely strong wind, as well as being cold and dry. The wind strength on the surface is often 35 to 40 miles per hour, but speeds of over 80 miles per hour have been recorded, and trains have been overturned; it is of special significance to aviation, as it was in the past to the navigation of sailing-ships. When the mistral is blowing, the sky is often cloud-less, but the wind frequently brings temperatures well below Freezing Point. Along the lower reaches of the Rhone, its power is displayed by the permanent set of the trees towards the south-east; gardens and orchards are protected from it by thick hedges of cypress, and many of the smaller houses have doors and windows only on the south-eastern side.

MIXED FARMING. The combination of *Agriculture* and *Pastoral Farming*.

MOCK MOON. See *Paraselene*.

MOCK SUN. See *Parhelion*.

MOFETTE. An opening in the ground from which carbon dioxide is emitted; it occurs in a region of former volcanic activity, as in the Yellowstone National Park, U.S.A.

MOLE. A jetty or *Breakwater* of large stones, rubble, or other material, laid in the sea so as to protect a harbour or anchorage from storms.

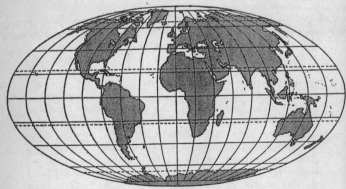

Mollweide's Projection.

MOLLWEIDE'S PROJECTION. A type of equal-area or *Homolographic Projection*, in which the entire earth's surface is depicted in one map. All parallels of latitude and the central meridian (usually the

Greenwich meridian) are straight lines, the latter being at right angles to the former, but other meridians are curved, and curvature increases towards the marginal meridians. Each area bounded by two adjacent parallels and two adjacent meridians is equal to any other area similarly enclosed. There is therefore no distortion of areas (as in Mercator's projection), and the projection is suitable for showing distribution of phenomena, etc., and generally speaking, for a study of world geography, but distorts shapes, directions, and distances, and is therefore useless for navigation.

MONADNOCK. An isolated hill or mass of rock which stands above the surrounding country because its rock has been more resistant to erosion than the rock of the latter: named after Mt Monadnock, in New Hampshire, U.S.A., which was formed in this way.

MONOCLINE. A *Fold* in which the bend is in only one direction; the rock stratum changes its *Dip* by increasing its steepness of inclination, and then levels out again or resumes its gentle dip.

MONSOON. The type of wind system in which there is a complete or almost complete reversal of prevailing direction from season to season. It is especially prominent within the tropics of the eastern sides of the great land masses, but it occurs, too, outside the tropics, for in eastern Asia it reaches as far north as about 60° N. Southeast Asia is pre-eminently a monsoon region. Here the intense heating of the land in the summer causes a low-pressure area to be established over NW. India, and the inblowing winds of the southwest monsoon, having originated as the SE. *Trade Winds* in the southern Indian Ocean, are warm and saturated with moisture. They advance northwards over the region from spring to midsummer, bringing copious rains to large areas, especially where mountain ranges enhance the effect, e.g. on the slopes of the mountains of western India and Burma, the eastern Himalayas, the Assam Hills, the Ganges and Irrawaddy deltas; the strong winds, heavy seas, overcast skies, and frequent rainfall contrast sharply with winter conditions. The south-west monsoon moves forward with a definite front, and arrives at each place at approximately the same date each year; the first rain is known as the 'burst of the monsoon'. Violent *Tropical Cyclones* occur in its van and rear and occasionally while it is at its height. The summer monsoon lasts from about April to September, though its duration anywhere depends upon the geographical situation of the place in question. In autumn, as the land cools, high pressure is established over the Asiatic continent, with lower pressure to the south. The north-east monsoon, which now advances southwards as the south-west monsoon retreats, is thus to most of the region a cold, dry wind, bringing rain only to the limited areas which it reaches by a partial sea-track, e.g. SE. India, Ceylon. The prevailing direction of this monsoon in N. China is NW. and in Central China N., and owing to the proximity of the deserts the winds are dusty as well as dry and cold; N. India is protected by the Himalayas and the Tibetan plateau, and the winds are much lighter than in China. The relief of the land largely controls the effect of the winter, as well as the summer, monsoon. It lasts approximately from October to March. Other areas which experience similar but less pronounced seasonal changes of wind direction include the south-eastern United States, East Africa, and north Australia, and these are sometimes rather misleadingly termed 'monsoon regions'. In the SE. United States there is not a complete reversal of wind; in the southern continents there are damp, inblowing winds on the east coasts during the summer but the extra-tropical land masses are too small to develop winter anticyclones of the Asiatic type, and there is no marked system of outblowing winds.

MONTAÑA. The mountainous region on the eastern slopes of the Andes, in South America, which has a heavy rainfall and is largely covered with dense forests, though not in such profusion as on the lowlands of the Amazon basin.

MONTH. See *Lunar Month, Calendar*.

MOON. The satellite of the earth, the only heavenly body which revolves round the earth; in astronomy the term is also applied, however, to the satellite of any planet. One revolution of the moon round the earth is completed approximately once each lunar month. Its sidereal period of revolution (i.e. with respect to the so-called 'fixed' stars), is about $27\frac{1}{4}$ days; its synodic period of revolution (i.e. the interval from new moon to new moon) is about $29\frac{1}{2}$ days, but varies by about 13 hours because of the eccentricities of the orbits of the moon and the earth. The diameter of the moon is rather more than one quarter of that of the earth. Unlike the sun and the stars, the moon shines only by reflected light, the light having been emitted by the sun, and only the half which is illuminated by the sun can be seen; as it revolves round the earth, this illuminated portion appears and disappears. As the moon rotates on its axis once in each orbital revolution, the same face of the moon is always turned towards the earth. See *Moon, Phases of*.

MOON, PHASES OF. The various stages or changes of appearance through which the moon passes during its revolution round the earth, known as new moon, full moon, etc., owing to the changes in relative position of the earth, moon, and sun. When the moon is between the earth and the sun, the dark side faces the earth, and no moon is visible; this period is called the *new moon*, for immediately afterwards a 'new' moon begins to appear. When the moon has revolved through about one-eighth of its orbit, its illuminated portion, as seen from the earth, has a crescent-shaped appearance. When about one-quarter of the orbit has been traversed, one-half of the moon's surface is seen; this is called the *first quarter*. After another eighth of the orbit, three-quarters of the surface is seen; this is the *gibbous moon*. Then, when the earth is between the moon and the sun, all of the moon's illuminated portion is visible; this is the *full moon*. Throughout these phases, from new moon to full moon, the moon is said to have 'waxed'; for the second half of its orbit, when the amount of moon visible from the earth decreases from full moon to new moon, it is said to 'wane'. It will be noted that new moon and full moon occur when earth, moon, and sun are practically in the same straight line, while first quarter and last quarter occur when the line from earth to moon is perpendicular to the line from earth to sun.

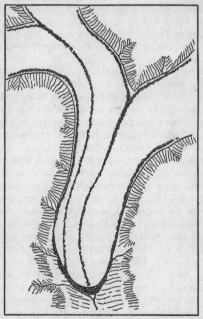

Diagram to illustrate the formation of lateral, medial, and terminal Moraines.

MOOR, MOORLAND. A wild stretch of land, usually elevated, and covered with heather, coarse grass, bracken, or similar vegetation;

it sometimes includes patches of pasture grasses, and sometimes
marshy hollows.

MORAINE. The debris or fragments of rock material brought down
with the movement of a *Glacier*. When lumps of rock, broken off by
frost and other agencies, fall down the valley slopes on to the ice,
they form the *lateral moraines*, one on each side of the glacier. When
two glaciers meet, two lateral moraines unite to form a *medial
moraine*; if other glaciers join them later, several medial moraines
may be formed. Fragments which are dragged along beneath the
ice form a *ground moraine*. At the end of the glacier, where the ice is
melting, all the debris which has been transported down is dropped,
and forms a *terminal moraine*, which extends across the valley. The
material of a moraine differs from that brought down by water, in
that large boulders may occur side by side with small fragments;
again, the lumps of rock, instead of being rounded, are angular
in shape and often scratched, for they are dragged and not rolled
along. (See Fig. p. 119.)

MORASS. A *Bog* or *Marsh*.

MORTLAKE. A term sometimes used, mainly in England, for an
Ox-Bow Lake.

MOULIN (French) or GLACIER MILL. A vertical shaft or pipe
formed in a glacier when streams of water from the melting ice
reach a *Crevasse* and plunge inside, carrying stones and rock waste
with them; the water takes on a rotary movement, and whirls its
load of stones round as it rushes through the ice, forming the moulin
or glacier mill. The whirlpool of water not only reaches the base of
the glacier, but acts upon the bed of rock beneath.

MOUNTAIN. A mass of land considerably higher than its surroundings,
and of greater altitude than a hill; an eminence is often considered a mountain rather than a hill when its elevation from foot
to summit is well over 1,000 ft, but the distinction is arbitrary. The
summit area of a mountain is small in proportion to the area of its
base; in this respect it differs from a *Plateau*, which might be of
similar elevation. Although the highest mountains are higher than
any plateau, many mountains are less elevated than the highest
plateaux – e.g. the Tibetan plateau, 15,000 to 16,000 ft above
sea level. Few of the highest mountains occur as single, isolated
peaks; most of them are arranged in ranges or groups. See *Range*,
Chain. Mountains may be formed by earth movements. See *Folded
Mountain, Block Mountain*. They may also be formed by *Erosion*,
the more resistant rocks being left while the softer rocks surrounding them are worn away. See *Circumdenudation* or *Circumerosion*,
Mountains of; *Relict Mountains*; *Monadnock*. Thirdly, mountains
may be formed by volcanic action; these are often known as

mountains of accumulation. See *Accumulation, Mountain of.*

MOUNTAIN CLIMATE. The type of climate which is chiefly determined by relief and altitude rather than by latitude and situation with respect to the sea; it may thus occur in any region, and is somewhat outside the scheme of the main *Climatic Regions.* Compared with the lowlands, mountains show a decrease in atmospheric pressure, temperature, and humidity, and increase in intensity of insolation and radiation, and up to a certain height a greater rainfall. To some extent the various climatic zones up the mountain slopes correspond to those which extend from the equator to the poles, and altitude so tempers heat in low latitudes that a high mountain peak, even on the equator, may be snow-covered throughout the year. But there are fundamental differences: the intensity of sunshine through the thin, clear atmosphere above the *Snowline,* for instance, is much greater than that received obliquely through a considerable thickness of atmosphere in the polar regions. There is often diurnal reversal of wind direction, corresponding to land and sea breezes. See *Anabatic, Katabatic, Mountain Winds.*

Owing to the establishment of temperature inversions during the colder months and at nights, temperatures may then be relatively high on mountain summits, and diurnal and annual ranges of temperature in middle and higher latitudes may thus be lessened. Ranges of temperature, however, are very considerable in valley bottoms, and are much greater on plateaux than on mountains of the same height. There are great contrasts of temperature, too, between sunny and shaded mountain slopes. See *Adret, Ubac.*

Variations of temperature and rainfall are great within a small area, and there are corresponding variations in natural vegetation, crops, and occupations; whatever the zone of natural vegetation at the base of the mountain, the region of maximum precipitation is almost always forest-covered, with coniferous forest above and deciduous forest below.

MOUNTAINS OF CIRCUMDENUDATION or CIRCUMEROSION. See *Circumdenudation* or *Circumerosion, Mountains of.*

MOUNTAIN SICKNESS. A complaint which affects people at high altitudes, and appears to be due to the diminution in the oxygen supply to the lungs. One of the first signs is the increase in the speed of respiration; and headaches, sleeplessness, and sickness may be caused. If the change in altitude is made gradually, no ill effects may be suffered. Muscular exertion may accelerate the mountain sickness, but does not seem to be an essential factor; on the other hand, the expenditure of muscular energy may be severely limited: the average time for which riveters could work on the Oroya Railway of Peru, for instance, was one week. The severity of the complaint varies with individuals, and those in good physical training are the less affected.

MOUNTAIN WIND or MOUNTAIN BREEZE. The flow of cool air at night down mountain slopes and valleys; it often alternates with the daytime *Valley Wind*. See *Katabatic Wind*.

MOUTH. See *River*.

MUD VOLCANO or PAINT POT. A mound of mud, sometimes conical in shape with a crater at the top, formed in a volcanic region by hot water issuing from the ground mixed with fine rock material. Mud volcanoes thus occur in New Zealand and Sicily. They are also found, however, at Baku, U.S.S.R., in southern Baluchistan, and in other non-volcanic regions; there the water is ejected by gases of other than volcanic origin – at Baku, for instance, by volatile gases from the petroleum wells.

MULATTO. The offspring of a white and a Negro.

MULGA SCRUB. The *Scrub*, consisting largely of a species of acacia known as mulga, which covers parts of western and central Australia.

MUSKEG. A bog or *Swamp* of the *Tundra* or the *Coniferous Forest* region of Canada.

N

NADIR. A hypothetical point in the heavens directly opposite, or antipodal, to the *Zenith*.

NAPPES (French). Huge recumbent *Folds* that have been pushed far from their original positions.

NATURAL BRIDGE. A bridge or arch of natural rock. Sometimes such a bridge is formed when the tough, unjointed rocks of a valley spur are cut through by the stream. It is more common, however, in *Limestone* country, where it is probably the remnant of the roof of an underground cave or tunnel.

NATURAL REGION. An area which possesses within its borders a comparatively high degree of uniformity of relief, climate, and vegetation, and therefore to some extent a uniformity in human activities. In some cases a natural region corresponds closely to a *Climatic Region*, or to an area in which a certain type of *Natural Vegetation* prevails.

NATURAL VEGETATION. The vegetation of a region as it exists or has existed before being modified by man, e.g. through agriculture.

NAUTICAL MILE. A unit of distance used in navigation, equal to the length of one minute (1′) of arc on a *Great Circle* drawn on a sphere having the same area as the earth; this is approximately 6,080 ft. For all ordinary purposes the length of one minute of arc along a meridian, the length of a minute of arc on the equator (the *Geographical Mile*), and the nautical mile may all be considered equal to 6,080 ft.

NAZE. A promontory or headland.

NEAP TIDES. The *Tides* of small amplitude produced when the gravitational pull of the sun opposes, i.e. is at right-angles to, that of the moon; a consideration of the phases of the moon will show that they occur twice a month, about the time of first and last quarters. See *Moon, Phases of*. At these times the sun tends to produce high water where the moon would produce low water, and vice versa. There is thus an unusually small difference between high water and low water; the high tide is lower and the low tide is higher than usual.

NEBULAR HYPOTHESIS. The hypothesis which proposes that all the matter of the solar system was once a nebula, a slowly rotating mass of hot gas which extended beyond the orbit of the most distant existing planets. This nebula cooled, shrank, and rotated faster, existing planets having been formed by breaking away from the central mass; it is now represented in its original form by the whitehot, central sun of the solar system. The cooling process caused the earth to form an atmosphere and a molten globe, which then cooled further so that a solid crust was formed; the water vapour in the atmosphere condensed to form water, and this fell as rain and produced the oceans.

NEHRUNG (German). Along the Baltic coast of East Germany and Poland, a *Spit* at the mouth of a river which forms a *Haff* or lagoon.

NEKTON. The larger, actively swimming marine creatures, principally fishes, as distinct from the passively drifting *Plankton* and the sedentary *Benthos*.

NEPHOSCOPE. An instrument used to measure the speed and direction of motion of the clouds; generally speaking, its use is confined to observations on medium and high clouds. One of the commonest types is the Besson comb nephoscope, which carries a number of pointed rods fixed like the teeth of a comb to the top of a long vertical rod; by rotating the rod till the clouds appear to move along the points, their direction can be determined, and if the height of the clouds be estimated, their speed can also be calculated.

NESS. A promontory or headland.

NEVADOS. The cold winds which blow down from the mountains with considerable regularity in the higher valleys of Ecuador; they are caused by the cooling of the air on the mountains, partly by nocturnal radiation, and partly by contact with ice and snow, the cold air flowing down the slopes. See *Katabatic*.

NÉVÉ (French) or FIRN (German). The granular substance, partly snow and partly ice, which is formed as the snow that has accumulated at the head of a glacier valley is being transformed into glacier ice. As the glacier moves down the valley, the névé gradually becomes welded into a completely crystalline mass of ice.

NIMBOSTRATUS. A type of low *Cloud*, dark grey in colour and of almost uniform base, which often gives continuous rain or snow; sometimes the precipitation does not reach the ground, and portions of the cloud then appear to trail.

NITROGEN. The most abundant gas in the *Atmosphere*, occupying approximately 78 per cent by volume of dry air; it is chemically inactive, and serves to dilute the *Oxygen*.

NIVATION. *Erosion* due to the action of snow. An example is the *Solifluction* or *Soil Creep* caused by the thawing of a thin snow cover and the upper parts of the ground, when the latter slides downwards over the frozen mass beneath.

NOMADISM. The practice among certain primitive peoples of frequently changing their habitation: especially applied to inhabitants of the *Steppes* and deserts, who wander in search of pastures for their livestock, or food such as fruit and roots for themselves, or, possibly, for trade. Such people are known as nomads. When food is scarce, pastoral nomads, such as the Bedouins, may be forced to raid their settled agricultural neighbours.

NORTE. The northerly wind experienced in winter in Central America, being a continuation of the United States *Norther*, and likewise bringing a sudden and often considerable fall of temperature; the temperature may be reduced by 6° to 9° C for a day or two – a severe change for such latitudes – and the wind may blow with gale force on exposed coasts. On the northern coasts of the Isthmus of Tehuantepec, where the nortes are strong and almost constant throughout the winter, these winds are accompanied by heavy rainfall; when they reach the Pacific side, on the Gulf of Tehuantepec, they are still cold but dry. The name is also applied to the cold, northerly wind experienced in eastern Spain during winter, when the high atmospheric pressure of the interior produces an outblowing current.

NORTHER. The *Cold Wave* sometimes experienced in the southern United States, usually in rear of a depression. The temperature may fall by about 20° C in 24 hours, causing great damage to fruit crops. The norther often blows with great violence, possibly 40 to 60 miles per hour, is accompanied by severe thunderstorms and hail, and may give *Line Squall* conditions. In the Sacramento valley of California, however, this wind is often dry and dusty, having been heated and dried by descent from the mountains.

NORTHERN CIRCUIT. The general northerly route by which depressions cross the North American continent from west to east, via the Great Lakes and the St Lawrence basin. It is the most frequented track in summer, though the depressions are then fewest in number, weakest, and the least constant. See *Southern Circuit*.

NORTHERN LIGHTS. See *Aurora*.

NORTH MAGNETIC POLE. See *Magnetism, Terrestrial*.

NORTH POLE. See *Poles*.

NOR'-WESTER. (1) In New Zealand, a hot, dry *Föhn* wind which descends from the mountains.

(2) On the plains of northern India, a type of *Squall*, usually accompanied by violent thunderstorms and heavy rain and hail showers, experienced during the hot season (April to June). At this season the rainfall of Bengal, Assam, and Burma is largely derived from these nor'-westers, and in Assam they are extremely important to the tea crop.

NUÉE ARDENTE (French) or PELÉAN CLOUD. The blast of hot, highly gas-charged, swiftly moving lava fragments sometimes ejected horizontally from a volcano when upward movement has been obstructed; despite the French term, the cloud surmounting the blast does not always glow. The English term is derived from Mont Pelée (see *Volcano*).

NULLAH. In India, a normally dry watercourse, which only temporarily becomes a stream immediately after heavy rain. See *Arroyo, Wadi*.

NUNATAK. An isolated mountain peak or hill projecting, like an island, from the mass of snow and ice near the margin of an *Ice-Cap*, e.g., in Greenland. The nunataks become more numerous towards the extreme edge of the ice-caps, for there the ice is thinnest.

O

OASIS.* An area in the midst of a desert, which is made fertile by the presence of water. It sometimes consists merely of a clump of palm-trees, sometimes it is a fertile region several hundred square miles in area. In the smaller type the water usually consists of a small isolated spring; alternatively it may be a spot where a stream flows down from a mountain canyon to form an *Alluvial Fan*. The larger oases, such as those formed by the Nile and the Euphrates, may support a huge, agricultural population. In the oases of the hot deserts, the typical tree is the date-palm, the fruit of which forms the main food supply.

OBSEQUENT RIVER. A tributary to a *Subsequent River*, flowing in the opposite direction to the original *Consequent River*.

OCCLUSION. In a *Depression*, the closing of the *Cold Front* on to the *Warm Front*. The *Warm Sector* of the depression is lifted by the advancing cold air, and at the earth's surface is reduced to a line – the line of the occlusion; it still exists in the upper air, however, and the cloud and rain of the original fronts remain with it for some time. On European *Weather Charts*, lines of occlusion appear quite frequently, for most of the depressions passing eastwards across the North Atlantic are occluded by the time they reach Western Europe.

OCEAN. The sheet of salt water which surrounds the great land masses of the earth; it is divided by them into several extensive portions, each known as an ocean, and altogether covers about 71 per cent of the earth's surface. The individual oceans are the Pacific, the largest, flanked by high mountain chains, having a deep, fairly level floor, and remarkable for the number of small islands, largely of volcanic origin; the narrower Atlantic, with few islands; the Indian, still smaller, and enclosed on three sides by continents and islands; the Arctic, surrounding the North Pole; and the Southern Ocean, surrounding Antarctica.

The oceans may be divided into three distinct regions, the *Littoral*, the *Pelagic*, and the *Abyssal*. Alternatively, they may be divided according to depth into four regions: (1) the *Continental Shelf*, which adjoins the land, (2) the *Continental Slope*, immediately outside the continental shelf, (3) the *Deep-Sea Plain*, a wide and almost level area forming most of the ocean floor, and varying in depth from about 2,000 to 3,000 fathoms, and (4) the *Deeps*. The ocean floor, like the land surface, has many ridges and valleys; its area and depth may be represented diagrammatically by a *Hypsographic Curve*. At the surface, the temperature of the ocean is usually a few degrees higher than that of the air above, and thus varies considerably, from over 80° near the equator to below Freezing Point in the polar regions. But ocean temperature decreases rapidly with depth below the surface; below 2,000 fathoms, temperatures are uniformly low in all oceans, being only a few degrees above the Freezing Point. The *Salinity* of the oceans also varies.

OCEAN CURRENT. A movement of the surface water of the ocean. Of the various causes of ocean currents, the most important are the prevailing winds and the differences in density due to variations in temperature or salinity; some deflexion of currents is caused by the rotation of the earth. Those currents caused by prevailing winds are known as drift currents, the best-known being the Gulf Stream. The circulation of the main ocean currents between the equatorial and the temperate regions is clockwise in the northern hemisphere and anti-clockwise in the southern hemisphere. In the Atlantic, for instance, the currents driven by the NE. and SE. *Trade Winds* combine near the equator to form the westward-moving equatorial current. This divides off the Brazilian coast, the northward portion forming the Gulf Stream, which moves along the east coast of the United States, meets the cold Labrador current off Newfoundland, and in the main continues with the prevailing *Westerlies* to Europe; then it divides, the northern branch reaching the coasts of NW. Europe as a warm current, passing into the Arctic Ocean, while the southern branch moves to the coast of NW. Africa as a cold current. (The terms 'warm' and 'cold' for a current denote their temperature relative to that of the neighbouring atmosphere.) The currents of the South Atlantic are very similar, though the general circulation is anti-clockwise. Those of the Pacific are also similar, the warm current of the North Pacific corresponding to the Gulf Stream of the North Atlantic being the Kuro Shiwo. Ocean currents often have a considerable influence on climate; the North Atlantic Drift, for instance, is largely responsible for the mild winters of NW. Europe, and the meeting of the Gulf Stream and the cold Labrador current gives the dense fogs off Newfoundland. Currents caused by differences of temperature are known as *Convection* currents. Warm water of the equatorial regions flows towards the polar regions, where it is cooled, sinks, and returns along the ocean bed, as a bottom current, towards the equator; it then rises to the surface again. An example of a current due to a difference of salinity is the surface current which enters the Mediterranean Sea from the Atlantic Ocean. Owing to a small rainfall, great evaporation, and the small number of inflowing rivers, the Mediterranean has a higher salinity than the Atlantic; a relatively fresh surface current thus flows from the Atlantic into the Mediterranean while a relatively salt undercurrent flows in the reverse direction.

OCEANOGRAPHY. The study of the oceans, including the nature of the water, its movements, its temperature, its depth, the ocean bed, the flora and fauna, etc.

OCTOROON. The offspring of a *Quadroon* and a white, thus having one-eighth Negro blood.

ONION WEATHERING. See *Exfoliation*.

OOZES. The deposits, mainly liquid mud, which cover the *Deep-Sea Plain* of the *Oceans*. They consist largely of the remains of various kinds of *Plankton*, which live in the surface waters, and, on dying, settle to the bottom; they also contain volcanic dust, which has been carried by the wind, and has likewise sunk to the ocean bed. Oozes are named according to the organisms which are most abundant in them. See *Pteropod Ooze*, *Globigerina Ooze*, *Diatom Ooze*, *Radiolarian Ooze*, and *Red Clay*.

OPACO. See *Ubac*.

OPPOSITION. The position of two heavenly bodies when they are in the opposite direction to each other as viewed from the earth, i.e. their longitudes differ by 180°. The term is usually applied to the position of a heavenly body when its direction, as viewed from the earth, is opposite to that of the sun.

ORBIT. The path of a heavenly body through space in relation to some selected point.

ORBIT, EARTH's. The path of the earth round the sun. This path is very nearly circular, actually elliptical; the sun is not at the centre, but at one of the foci of this ellipse. The earth is thus nearer to the sun at one season than at the opposite season, when it has reached the farther side of the ellipse. See *Aphelion, Perihelion*.

ORDNANCE SURVEY. An accurate and detailed geographical survey made for the Government, originally by the War Office.

ORDNANCE SURVEY DATUM. See *Mean Sea Level*.

ORE. A mineral aggregate which is worth mining on account of the valuable metalliferous *Minerals* it contains.

OROGENESIS. Mountain-building: changes in the level of the earth's crust in which rocks are thrown up into folds or blocks to form ranges of mountains. In course of time the mountains may be considerably denuded, and may thereafter be depressed or elevated by *Epeirogenesis*.

OROGRAPHIC RAIN. Rain which is caused by mountains standing in the path of moisture-laden air; this air is forced to rise, is thereby cooled, and, if sufficient water vapour is present, rain is deposited on the high ground. Substantial amounts of orographic rain are produced by mountain ranges rather than by isolated peaks, especially when they are aligned at right-angles, or nearly so, to the direction of the wind. On account of orographic rain, places in mountainous or hilly regions often receive a consistently higher rainfall than neighbouring places situated on low-lying ground. See *Rainfall*.

ORTHOGRAPHIC PROJECTION. The type of *Map Projection* in which the globe is viewed as if from an infinite distance, so that only one hemisphere is depicted; the projecting rays are parallel and perpendicular to the plane of projection. At the margins there is considerable compression, and at the centre expansion, and distances, directions, shapes, and areas are distorted. The projection, which is a kind of *Zenithal Projection*, is of little geographical value, but is frequently used for charts of the heavens.

ORTHOMORPHIC PROJECTION. A type of *Map Projection* in which the *shape* of any very small area on the map is the same as the *shape* of the corresponding small area on the earth.

OUTCROP. The portion of a rock stratum which projects above the earth's surface, and is thus exposed to view.

OUTLIER. A mass of comparatively new rocks surrounded by older rocks. See *Inlier*.

OUTPORT. A port which is auxiliary to another seaport, but, being situated nearer the sea, is more accessible than the latter to the larger vessels: e.g. Avonmouth is an outport of Bristol, Cuxhaven of Hamburg.

OUTWASH PLAIN or SANDR. The alluvial plain formed by the streams originating from the melting ice of a glacier, which carry away some of the material of the moraine, and deposit it over a considerable area; the coarser material is deposited near to the ice, the finer material farther away.

OVERFOLD. A *Fold* which is overturned: i.e. except immediately adjacent to the apex, the *Dip* is in the same direction on both sides of the axis.

OX-BOW LAKE OR CUT-OFF. A lake formed when a meandering river (see *Meander*), having bent in almost a complete circle, cuts across the narrow neck of land between the two stretches, and leaves a backwater; silt is gradually deposited by the river at the entrances to this backwater, till the latter is finally separated from the river, and becomes a lake. Such ox-bow lakes are very common beside the banks of the Mississippi, and are often known as *Bayous*.

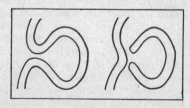

Diagrams of an exaggerated meander and an Ox-Bow Lake.

OXYGEN. The chief gas in the *Atmosphere* from the viewpoint of mankind and the animals, being essential to the process of breathing: in the body it unites with carbon to form *Carbon Dioxide*, which is then exhaled. Oxygen occupies about 21 per cent by volume of dry air. It is also the most abundant element, forming about eight-ninths by weight of water, and nearly one-half by weight of all the rocks of which the earth's crust is comprised.

P

PACK-ICE. Large blocks of ice, of greater extent and depth than *Ice-Floes*, which have been formed on the surface of the sea when the *Ice-Field* has been broken up by winds and waves, and have drifted from their original position. They are termed 'close pack' if the blocks are mainly in contact, 'open pack' if they are not. Most of the pack-ice gradually melts and disappears in warm weather.

PAHOEHOE. The Hawaiian term for *Ropy or Corded Lava*, often used in place of its English equivalent.

PAINT POT. See *Mud Volcano*.

PALAEOGEOGRAPHY. The study of the distribution of land and water, etc., during earlier periods of the earth's history.

PALAEONTOLOGY. The science which deals with extinct organisms, of either animal or plant origin, whose remains are found buried in the rocks; the study of *Fossils*. The term is now often applied only to the study of animal fossils, the study of plant fossils being termed palaeobotany.

PALAEOZOIC ERA. The Ancient Life Era, the second of the five major sub-divisions of the geological scale of time. Its rock strata are the oldest in the earth's crust from which definite organic remains have been recorded, though during the era the creatures inhabiting the earth were very different from those of the present day. For long the vertebrates were represented only by fish, but later amphibians appeared, and, at the end of the period, reptiles. The vegetation consisted of cryptogamous plants, such as ferns and mosses. Late in the era, layers of partially decayed vegetation became buried under marine deposits, and by compression and hardening formed coal. See *Carboniferous Period*. It is sometimes known as the Primary Era.

PAMPAS. The mid-latitude *Grasslands* or grassy plains lying round the River Plate estuary in South America, from the Andes to the Atlantic Ocean. The western part of the pampas is largely desert, but the eastern part, which has a higher rainfall, has a natural covering of tall, coarse grass, known as pampas grass. Vast areas of the pampas are now cultivated, yielding wheat, maize, alfalfa, and flax (for linseed), and enormous numbers of cattle and sheep are reared for the frozen meat industry. In physical characteristics, the pampas are similar to the *Prairies* of North America, the *Steppes* of the U.S.S.R., and the *Veld* of South Africa.

PAMPERO. A violent burst of cold, polar air experienced on the *Pampas* of Argentina and Uruguay after a depression has passed. It blows from a southerly to westerly direction, and takes the form of a *Line Squall*, with the typical roll of cloud; it is sometimes accompanied by rain, thunder, and lightning, and dust is swept up from the pampas. A considerable fall of temperature occurs as the storm passes. The pampero is most frequent in summer, and is comparable with the *Southerly Burster* of Australia.

PANCAKE ICE. The small, thin cakes of ice, which form on the surface of the water when the sea in polar regions begins to freeze.

PAPAGAYO. The cold, northerly wind sometimes felt on the Mexican plateau, being a continuation of the *Norther* of the United States, and comparable with the *Norte* of the coastal region.

PARALLEL OF LATITUDE. See *Latitude*.

PARAMOS. The bleak, barren lands extending above the *Puna* of the Andes, in South America, as far as the snowline, and invaded in parts by glaciers. Although flowering plants grow in sheltered spots, the vegetation over much of the area is stunted, as on the *Tundra*, and consists largely of lichens and mosses.

PARASELENE or MOCK MOON. An image of the moon, similar to the *Parhelion* or mock sun; it is formed at the same angular distance from the moon, 22°, as the parhelion is from the sun, when the moon is near the horizon, and at a greater distance when the moon has a higher elevation.

PARHELION or MOCK SUN. An image of the sun, usually coloured, with the red nearest to the sun, and at the same elevation as the sun. When the sun is near the horizon, the angular distance of the mock sun from the sun equals that of the ordinary *Halo*, 22°, but when the sun is higher the distance is greater, and the mock sun is outside the halo. Mock suns are sometimes white.

PARTIAL DROUGHT. In the British Isles, a period of at least twenty-nine consecutive days, during which the mean daily rainfall does not exceed .01 in. The definition has not been internationally accepted.

PASS. A low and passable gap through a mountain barrier. It is usually caused by erosion, either by a glacier or by two streams rising close to each other on opposite sides of the mountain barrier. Being approached from both sides of the mountains by a steep valley, and being flanked by mountain peaks, the pass is thus saddle-shaped.

PASTORAL FARMING. The practice of breeding and rearing certain herbivorous animals. By its means it is possible for man to satisfy almost all his needs in food, clothing, and shelter; in some regions, in fact, it is essential for him to do so. Some inhabitants of the tundra, for instance, derive most of their necessaries of life from the reindeer, those of desert or semi-desert lands from sheep or goats. The flocks and herds sometimes suffer from drought, cold, or disease, but when large numbers are reared, it is very rare for all the animals to be destroyed. The need for pastures for the animals causes the inhabitants of relatively inhospitable regions to lead a nomadic life; on most of the *Grasslands* of the world, however, the pastoral farmers are now following a settled existence.

PEAK. The top of a mountain or hill, standing above the level of the range or the surrounding country. It is usually formed through the *Erosion* of adjacent rocks, those of which the peak is composed being of more resistant material. Sometimes, however, it may be the result of a *Fold* or a *Fault*, or it may be a volcanic cone.

PEAT. A brownish or blackish fibrous substance produced by the decay of vegetation and found in *Bogs*; it is usually considered to represent the first stage in the transformation of vegetable matter into coal, and contains a high proportion of water. See *Lignite*.

PEDALFER. One of the two types into which soils are sometimes classified; it is roughly equivalent to 'soil of a humid region', and, as the name suggests, is generally rich in iron and clay. The pedalfers are divided into podzolic and laterite soils (see *Podzol, Laterite*). See *Pedocal*.

PEDIMENT. In arid and semi-arid regions, the gently sloping plain strewn with boulders that borders the mountains.

PEDOCAL. One of the two types into which soils are sometimes classified; it is roughly equivalent to 'soil of an arid region', and, as the name suggests, is generally rich in calcium. The pedocals are divided into pedocals of the temperate zone and of the tropical zone. See *Pedalfer*.

PEDOLOGY. The science of the study of soils.

PELAGIC. Belonging to the open sea, i.e. to that part of the *Ocean* which is deeper than the *Littoral* region but shallower than the *Abyssal* region.

PELAGIC DEPOSITS. The deposits which cover the bed of the *Pelagic* zone of the ocean, formed largely of the remains of plants and animals of microscopic size which in life float in the surface waters; they include, however, the products of volcanic eruptions. See *Oozes*. There is, in fact, no sharp line of demarcation between pelagic deposits and *Terrigenous Deposits*.

PELÉAN CLOUD. See *Nuée Ardente*.

PENEPLAIN or PENEPLANE. A region which is almost a plain. Its formation is often due to erosion by rivers and rain, which continues until almost all the elevated portions are worn down; the more resistant rocks frequently stand above the general level of the land, though in time they, too, will be brought down to that level. See *Monadnock*. When a peneplain is raised, it often becomes a *Plateau*, which is then dissected by the rivers as they pass through a fresh cycle from youth to old age.

PENINSULA. A stretch of land almost surrounded by water.

PENUMBRA. The partly shaded region, from which a small amount of light is received, surrounding the *Umbra* in an *Eclipse*. The term is also applied, for instance, to the similar region surrounding the umbra of a *Sunspot*.

PERCHED BLOCK. An *Erratic Block* that has been left standing in an exposed and precarious position.

PERIGEE. The point in the orbit of the moon or of a planet, or in the apparent orbit of the sun, when it is nearest to the earth.

PERIHELION. The position of the earth, or of another heavenly body, in its *Orbit* when it is at its nearest point to the sun. The earth reaches its perihelion during the northern winter, i.e. at the beginning of January; although the earth is then nearest to the sun, the northern days are short, and the sun's rays fall very obliquely on the northern hemisphere, and the decreased insolation more than counterbalances the shorter distance. See *Aphelion*.

PERMAFROST. Ground that is permanently frozen, as in polar regions.

PERMEABLE ROCKS. Rocks which, being porous, allow water, e.g. rain water, to soak into them; sandstone is an example. Some geographers assume the term to be synonymous with *Pervious Rocks*.

PERVIOUS ROCKS. Rocks which allow water, e.g. rain water, to pass through them freely, being either porous or fissured. As the water can only pass through the rocks gradually, an exceptionally heavy fall of rain will cause a surface *Run-Off* even from these rocks. Some geographers use the term *Permeable Rocks*.

PETROLOGY. The study of the composition, structure, and history of the rocks forming the *Lithosphere*, or earth's crust, and their mineral structure; it is thus a branch of *Geology*.

PHACOLITH or PHACOLITE. A long, lens-shaped strip of *Igneous Rock*, placed near the top of an *Anticline* or the bottom of a *Syncline*. It is formed in a similar way to a *Laccolith*, by the intrusion of *Magma* from below without reaching the surface, but it is, of course, much shallower than a laccolith. See *Saddle-Reef*.

PHASES OF THE MOON. See *Moon, Phases of*.

PHENOLOGY. The science which deals with the effects of seasonal changes upon animal and vegetable life. It thus includes such phenomena as the flowering or bursting into leaf of plants, seed-time and harvest, etc., though observations are often limited to the dates at which these phenomena, and, for instance, the first and last appearance of birds and insects, occur. The relations between these dates and the latitude, longitude, and altitude of the various regions are also examined.

PHOTO RELIEF. The process by which a relief model of an area is photographed for reproduction on a map, called a photo relief map; the method merely depicts the salient physical features of the area.

PHYSIOGRAPHY. The study of the physical features of the earth, their causes, and their relation to one another. It is sometimes held to be synonymous with the more modern term *Geomorphology*, and sometimes, rather loosely, with Physical *Geography*.

PHYTOGEOGRAPHY. The study of the distribution of plants on the earth in relation to their geographical environment.

PIDGIN ENGLISH. A corrupt form of English used as a medium of communication between English-speaking traders and various native races. It was originally the language of business between Englishmen and Chinese, and the words are believed to be a Chinese corruption of 'Business English'. But different forms of Pidgin English are current among millions of natives of Africa, south-east Asia, Australia, and the Pacific Islands; each form consists of a few hundred English words and a much smaller number of common native words. See *Lingua Franca*.

PIEDMONT ALLUVIAL PLAIN. See *Alluvial Fan*.

PIEDMONT GLACIER. An extensive sheet of ice covering low-lying ground at the foot of a mountain range, formed by the union of several *Glaciers*. Its width is usually greater than the combined width of the glaciers which form it, and the rate of movement is therefore slow; sometimes, in fact, movement practically ceases. One of the best-known examples of a piedmont glacier is the Malaspina glacier in Alaska, which is 60 to 70 miles long and 20 to 25 miles wide.

PILLOW LAVA. A *Lava* which solidified under water, either because it was ejected under water or because it flowed into the water before solidification, assuming shapes that resemble a heap of pillows.

PILOT BALLOON. A small free rubber balloon, filled with hydrogen, which is used to obtain the direction and velocity of the upper winds. Usually the balloon is followed, after release, by means of a single *Theodolite*; as the balloon contains a known amount of hydrogen, it has a known rate of ascent, and by determining its position at fixed intervals, normally every minute, its direction and speed – and therefore those of the wind – can be calculated.

PIPE. In chalk country, a *Sink Hole* that has filled with gravel. These pipes may be seen in section in many chalk pits.

PIRACY, RIVER. See *River Capture*.

PITCH. The inclination to the horizontal of the axis of a *Fold*, i.e. the *Dip* of the beds along the axis.

PLAIN. An extensive area of level or gently undulating land, usually of low altitude. Plains may be formed in a variety of ways, and are often named accordingly: e.g. *Alluvial Plain, Coastal Plain, Flood-Plain*. On many of the plains of the temperate zone, the natural vegetation is grass, and the plains have thus come to be known as *Grasslands*, and, according to their location, as *Prairies, Pampas,* and *Steppes*.

PLANE OF THE ECLIPTIC. See *Ecliptic*.

PLANET. One of the spherical bodies, of which there are several hundreds, which revolve round the sun in elliptical paths known as their orbits. There are nine planets, including the earth, much larger than the rest; those larger than the earth are Neptune, Uranus, Saturn, and Jupiter, those smaller are Venus, Mars, and Mercury. Venus and Mercury are nearer to the sun than is the earth, the rest are farther away. Unlike the sun and stars, and like the moon, the planets shine only by reflected sunlight.

PLANE TABLE. A simple surveying instrument by means of which a sketch-map of a small area may be drawn; it consists of a board standing horizontally on a tripod, and carrying a ruler, a compass, and sights. A line representing magnetic north and south is drawn on the paper which covers the plane table, the directions of the various physical features are found by sighting, and distances are measured.

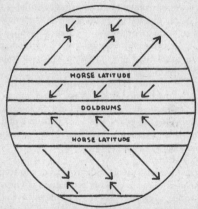

Diagram to show the Planetary Winds at the earth's surface.

PLANETARY WINDS. The general distribution of winds throughout the lower atmosphere, which – determined by differences in *Insolation* – would be set up similarly in any rotating planet possessing an atmosphere. It consists of the following: the calms and very light winds of the equatorial low-pressure belt – the *Doldrums*; the *Trade Winds*, blowing from the sub-tropical belts of high pressure towards the equator, from the NE. in the northern hemisphere and from the SE. in the southern hemisphere; the *Westerlies*, on the poleward side of the Trade Winds, being mainly SW. in the northern hemisphere and NW. in the southern hemisphere; and the outblowing polar winds, from the NE. in the northern hemisphere and the SE. in the southern hemisphere. The whole system of planetary winds and pressure belts moves north and south with the equatorial belt with which it is primarily associated.

PLANETESIMAL HYPOTHESIS. The hypothesis which proposes that the planets of the solar system were formed by the coalescence of large numbers of minute planets, or planetesimals, owing to collision and gravitational attraction. The idea derives from the fact that there are several spiral nebulae now existing in the universe, each one having a central nucleus about which opposite curved arms revolve; in the arms are dense spots, smaller than the nucleus, about which the planetesimals revolve. Thus the earth was produced from one nucleus, the moon from another. The result of pressure, as more and more material accumulated, was to heat the interiors of the planets, and this heat is now being lost again.

PLANETOID. See *Asteroid*.

PLANIMETER. An instrument for measuring irregular plane areas on charts and maps. The tracing point of the instrument is placed at a certain point on the perimeter of the area to be measured, and is then moved round the perimeter till the point is again reached. An index on the instrument gives the area which corresponds to the distance described by the tracing point.

PLANKTON. The minute, floating organisms of plant and animal origin, usually invisible to the naked eye, which inhabit the waters of seas, lakes, streams, and ponds, and form the food of many fishes and other creatures. The principal fishing-grounds of the world are thus found on the continental shelves, where plankton abound. Animal and vegetable plankton also form the food of the whale. See *Benthos, Nekton*.

PLANTATION. An estate, usually in a tropical or sub-tropical region, which is devoted to the large-scale production of one or more *Cash Crops*, e.g. coffee, sugar-cane. In the past the owners of plantations often employed slave labour.

PLATEAU. An extensive, level or mainly level area of elevated land. Sometimes the plateau is traversed by rivers and mountain ranges, as in the case of the *Meseta* of Spain; it may be exceptionally high, as in the case of the Tibetan plateau. If formed of horizontal strata, when young it will be intersected by deep *Canyons*; later, the valley floors widen, and it is broken up into *Mesas*; later still, most of the surface becomes a plain, dotted with *Buttes*. It is usually called a *Tableland* when it is bordered by steep, cliff-like faces or *Scarps*. Plateaux may be classified as intermontane, piedmont, or continental plateaux (or tablelands). The *intermontane plateaux* are those which have been formed in association with mountains, and are partly or completely enclosed by them; the highest and some of the most extensive plateaux, e.g. the Tibetan plateau, are of this kind. The *piedmont plateaux* lie between the mountains and the plains or the ocean, e.g. the plateau of Patagonia. The *continental plateaux* or *tablelands* rise abruptly from the lowlands or the sea, e.g. the plateau or tableland of South Africa. Many of the plateaux of the world have an arid climate.

PLAYA. A tract of land, such as an *Alkali Flat*, which is temporarily filled with water and becomes a shallow, muddy lake after exceptionally heavy rainfall, or the flooding of a river, but dries up again in hot weather. The Great Basin of the western United States, in Nevada and Utah, has many playas.

PLUM RAINS or BAI-U. The early summer rainfall of Japan, caused by depressions of continental origin passing eastwards over central China and across the Yellow Sea. During this period the rain falls very frequently, the sky is overcast, and the atmosphere is extremely humid; the large cloud amounts have the effect of checking the summer rise of temperature. The name is derived from the fact that the rains come at the time of ripening of the plums. This is the period when the rice plants are being transplanted, and the 'plum rains' are therefore extremely important to the rice crop.

PLUTONIC or ABYSSAL ROCKS. The *Igneous Rocks* which have solidified deep down in the earth, where the cooling has been slow and the various minerals have had time to crystallize; these rocks are thus always completely crystalline. Granite provides a good example.

PLUVIAL. Relating to, due to, or accompanied by rain.

PLUVIOMETRIC COEFFICIENT. The mean amount of rainfall at a place for a given period, expressed as a percentage of the 'normal', the 'normal' being the amount that would be precipitated if the rainfall were evenly distributed throughout the year.

PODZOL or PODSOL. A type of soil which is characteristic of regions having a sub-polar climate, and one which therefore develops mainly under a natural vegetation of *Coniferous Forests*. It develops, in fact, where the precipitation is considerable but evaporation limited; the upper layers are leached, and the soil has a characteristic greyish-white colour. On the whole, the podzols make poor agricultural soils. Huge areas of the coniferous forest regions of northern Canada and the northern U.S.S.R. are covered with podzols.

POINT. (1) A cape.

(2) One of the thirty-two directions or dividing lines of the compass.

POLAR WINDS. The extremely cold winds which blow from the area of high pressure around the poles towards the temperate regions. In the northern hemisphere they blow in general from the north-east, in the southern hemisphere from the south-east.

POLDER. In Holland, an area of land which has been reclaimed from the sea or from a lake. The land is surrounded by dykes to protect it against encroachment, and is then drained by pumping the water into canals which run along embankments and serve also as means of communication. The polders are usually very fertile, and form valuable agricultural land or pasture land for cattle.

POLES. The two points at the northern and southern extremities of the earth, known as the North Pole and the South Pole, which form the ends of the earth's *Axis*, and thus remain stationary while every other point on the earth's surface is rotating about the axis. The term is also used of the North and South Magnetic Poles (see *Magnetism, Terrestrial*).

POLE STAR. The star which is seen in the zenith at the North Pole, and therefore may be used to find true north from any point on the earth's surface where it may be seen, i.e. in the northern hemisphere. At any place, the height of the Pole Star above the horizon is equal to the latitude; at the equator, the Pole Star has no altitude, and is just visible on the horizon. To find the latitude of a place in the northern hemisphere, then, the height of the Pole Star above the northern point of the horizon may be observed. To find the Pole Star, an imaginary line may be drawn through the two stars known as the 'pointers' in the constellation called the Plough, and then produced to about five times its length; this line will very nearly pass through the Pole Star. In the Southern hemisphere, although the stars there visible may be used to find directions, there is no single star immediately above the South Pole. See *Southern Cross*.

POLJE. A closed hollow in a *Karst Region*, similar to but larger than a *Dolina*, being flat-bottomed and several miles in length. After heavy rain it may temporarily become a lake. It usually has a covering of soil, making cultivation possible (see *Terra Rossa*). It may be due to erosion, having been formed primarily by solution of limestone, or it may be of tectonic origin, having been initiated by earth movements; the former kind is sometimes known as a *Blind Valley*.

POLYCONIC PROJECTION. A modified *Conical Projection*, in which the distances between the meridians along every parallel of latitude are made equal to those distances on the globe. The name is derived from the fact that each parallel is thus constructed as if it were the chosen standard parallel for a simple conical projection. The central meridian is divided truly, each parallel is divided truly, and the meridians are formed by drawing curves through the corresponding points on successive parallels. This type of projection is far from equal-area or orthomorphic, and so is unsuitable for maps covering a large area.

PONENTE. A westerly wind which blows in the Mediterranean area.

PONOR. In a *Karst Region*, an open vertical shaft enlarged by solution of the limestone, leading downwards from a *Sink Hole*.

POROROCA. The tidal wave or *Bore* of the Amazon River.

PORT. A town possessing a *Harbour*, at which ships call to load and unload goods. One of the simplest types is the fishing *port*, while those which are used by the largest ocean-going vessels are often known as *seaports*. A *river port* is often situated at the highest point on the river which can be reached by vessels large enough to be used for trade. See *Airport*.

POT-HOLE. A hole worn in the solid rock, usually at the foot of a waterfall, by the revolutions of a stone; the latter has been kept in motion by the eddies of the swift-flowing stream. The term is also sometimes applied to the type of *Sink Hole* or *Swallow-Hole* commonly found in limestone districts.

PRAIRIES.* The gently undulating, almost flat, generally treeless, grassy plains – the mid-latitude *Grasslands* – of North America, covering the southern regions of Alberta, Saskatchewan, and Manitoba in Canada and the central United States from the foothills of the Rocky Mountains about as far east as the longitude of Lake Michigan. The light summer rains, with local droughts and high summer temperatures, permit a rich growth of natural grasses, but not, in general, trees; these conditions are also suited to the growth of cereals, and the prairies have become one of the most important areas of wheat production in the world. The prairies form the North American counterpart of the *Pampas* of South America, the *Steppes* of Eurasia, and the *Veld* of South Africa.

PRE-CAMBRIAN or **ARCHAEAN ERA.** The whole period preceding the *Palaeozoic Era*, and the first of the five great eras of geological time. The Pre-Cambrian rocks are now exposed over about one-fifth of the earth's land surface, but our knowledge of the life of the Pre-Cambrian world is still very slight. The rocks are of great economic importance, for they contain the world's richest gold mines, including those of the Transvaal.

PRECIPITATION. In meteorology, the deposits of water, in either liquid or solid form, which reach the earth from the atmosphere. It therefore includes not only rain, but also sleet, snow, and hail, which fall from the clouds, and dew and hoar frost. The total depth of liquid deposited – the solid forms being melted – is measured at meteorological and climatological stations by means of a *Rain Gauge*, and expressed in millimetres or inches. *Rainfall* is often used synonymously with precipitation to denote the total amount of water in all forms which is deposited on a given area.

PRESSURE, ATMOSPHERIC. See *Atmospheric Pressure.*

PRESSURE GRADIENT. The rate at which the atmospheric pressure changes horizontally in a certain direction on the earth's surface, as indicated by the position of the *Isobars* on a *Weather Chart*; when the isobars are close together, a rapid change of pressure from point to point at right-angles to the isobars is shown, and the gradient is steep, while when they are far apart the gradient is slight. The isobars thus bear a relation to the pressure gradient similar to that of the *Contours* on a map to the height gradient.

PRESSURE TENDENCY. The change in atmospheric pressure which has taken place at a given spot during a certain period, usually three hours, preceding a meteorological observation. A number of pressure tendencies from various points in an area is valuable in showing the regions where the pressure is in process of rising or falling. See *Isallobar*.

PREVAILING WIND. The wind, indicated by direction, at a certain place or in a certain area which has a considerably higher frequency than any other.

PROFILE, RIVER. A section showing the slope of a river from its source to its mouth; initially it is often irregular, with rapids and waterfalls in the upper reaches, but later it is much smoother.

PROFILE, SOIL. A section through the soil showing the different horizons, or layers, usually designated by the letters, A, B, and C, which extend downwards from the surface to the parent material.

PROMONTORY. A headland; a cliff or crag projecting into the sea.

PROTECTORATE. A term which is used very loosely to denote the relations between two states, one of which exercises some form of control over the other. It varies between suzerainty and a mere agreement on the part of the superior state to protect the safety of the inferior state, but a common feature of all protectorates is the prohibition of foreign relations for the latter, except those permitted by the former. The term is also applied to the protected state itself.

PSYCHROMETER. An instrument designed to measure the *Humidity* of the atmosphere; a *Hygrometer*. One form, Assmann's Psychrometer, is very similar to the ordinary hygrometer, consisting of wet and dry bulb thermometers, past which a current of air is drawn by means of a small fan operated by clockwork; the temperatures registered by both thermometers are thus given under the same conditions, and the results are more reliable than with the ordinary hygrometer.

PTEROPOD OOZE. A calcareous *Ooze* which consists chiefly of the shells of the floating molluscs known as pteropods, and covers relatively limited areas of the ocean bed. It occurs mainly on the ridges and plateaux which rise from the *Deep-Sea Plain*. Below about 1,500 fathoms, the shells of the pteropods dissolve in the sea-water, and the ooze is thus limited to depths of about 500 to 1,500 fathoms. At these depths it occurs off certain tropical islands, mainly in the Atlantic Ocean, e.g. round the Azores, to the west of the Canary Islands, etc.

PUMICE. A type of volcanic rock which was made extremely light and porous owing to the sudden release of steam and gases as it was solidifying, the material swelling up into a froth. Pumice will float on water, and it occurs in all the deposits which lie on the bed of the deepest parts of the oceans, especially in *Red Clay*.

PUNA. The higher and bleaker parts of the plateau of the Andes, lying between about 10,000 and 13,000 ft above sea level. Here the climatic conditions of the *Tierra Fria* are exaggerated, and temperatures tend to extremes. Even in summer the nights are bitterly cold; in winter the wind is so cold that it sears the skin, penetrates almost any type of clothing, and causes a number of diseases, especially of the lungs, being known in parts of Bolivia as the 'Harvest of Death'. Although the sunshine may be intensely bright and hot, in the shade the air is frigid, and is avoided by the inhabitants. Heavy shawls and ponchos, or cloaks, are worn to protect the body, and woollen masks to protect the face. Except for the three months when rain occurs, the lakes and rivers mainly consist of shallow *Salt-Pans* and dry watercourses. Only the hardiest of crops, such as barley and potatoes, can be grown, and wool-bearing animals such as the llama and the alpaca are raised on the coarse grass.

PURGA. The cold, north-easterly wind of the U.S.S.R., often bearing snow, especially so-called in the *Tundra*; it is the same wind as the *Buran*.

PUSZTAS. The flat, almost treeless grasslands or *Steppes* of the Hungarian Plains.

PUY (French). A term used in Auvergne, France, for a hill which is the cone of an extinct volcano, with or without associated lava; there are many examples in that region, but they also occur in other parts of the world. Puys may occur as isolated hills or in groups.

PYGMIES. People of exceptionally small stature, in general those whose average height is below five feet. Some of the best-known examples are certain tribes of the *Equatorial Forests* of Central Africa who live within a few degrees of the equator, and are known as Negrillos; they have an average height of about 4 ft 6 in., and have a dark yellowish skin. The Asiatic pygmies, inhabiting parts of Malaya, Sumatra, the Philippines, New Guinea, and other regions of south-east Asia, are known as Negritos.

PYRHELIOMETER. An instrument for measuring the *Radiation* from the sun.

Q

QUADRANT. An instrument in the shape of a graduated quarter-circle, used in astronomy and navigation particularly for measuring angles and altitudes.

QUADRATURE. The position of a heavenly body when, as viewed from the earth, it makes a right-angle with the direction of the sun; the term is usually applied to the moon, and thus coincides with first quarter and last quarter. See *Moon, Phases of*.

QUADROON. The offspring of a *Mulatto* and a white, thus having one-quarter Negro blood.

QUAGMIRE. An area of soft, wet ground, which shakes or yields underfoot; a *Marsh*; a Quaking *Bog*.

QUAKING BOG. See *Bog*.

QUARTZ. The commonest mineral, being present in many rocks and soils in a wide variety of forms, and often filling veins, joints, and cavities; it often contains admixtures of other minerals, including *Ores*. It consists of silica, a compound of silicon and oxygen. It forms the major proportion of most sands. In the pure state quartz is glassy in appearance, and is so hard that it will scratch glass, but it is brittle.

QUARTZITE. An extremely hard, resistant rock, formed from *Sandstone* by the filling of the pore spaces with *Quartz*.

QUATERNARY ERA. The modern Era, the last of the five major sub-divisions in the geological scale of time. Some geologists, however, regard it as simply a sub-division of the *Cainozoic Era*.

QUICKSAND. A mass of loose or unstable sand at the mouth of a river, on the sea coast, etc.

R

RACE. A rapid *Tidal Current*, usually through a narrow channel, often caused by the *Tidal Range* at one end of the channel being greater than that at the other. A race may also occur near a headland separating two bays.

RADIATION. The process by which a body emits radiant energy, e.g. in the form of heat; it causes a loss of heat, and therefore cooling. Radiant energy is constantly emitted in all directions by the sun, some of this reaching the earth and being converted into heat; the earth is constantly losing heat into space by radiation. During the day, the heat received from the sun by *Insolation* exceeds the amount lost by radiation, and the temperature rises till a maximum is reached. At night the reverse is the case, and the temperature falls till a minimum is reached. In summer, when the days are long and the sun is high, the ground grows gradually warmer, but in winter, when the days are short and the sun is low, radiation so far exceeds insolation that the ground becomes cold. The land loses heat by radiation more rapidly than water, and high ground more rapidly than low ground.

RADIOLARIAN OOZE. A type of *Ooze*, which is found on the ocean bed only in deep, tropical seas. It is a variety of *Red Clay*, from which it is distinguished by the presence of many siliceous organic remains, especially the shells of the small, simple animals known as radiolaria; it is therefore named after these. It covers considerable areas of the Pacific Ocean, particularly between 5° and 15° N., and a limited area in the Indian Ocean.

RADIO SONDE. An apparatus consisting of a free balloon filled with hydrogen, to which is attached a self-recording instrument for measuring atmospheric pressure and temperature and a radio transmitter. The apparatus is released, and the readings of pressure and temperature at various heights are transmitted to the ground station by the radio transmitter. The apparatus can be used in all weather conditions and to considerable altitudes.

RAILWAY GAUGE. See *Gauge, Railway.*

RAIN. Separate water drops which fall to the earth from the clouds, having been formed by the *Condensation* of water vapour in the atmosphere. Condensation of this water vapour is brought about by the rise of air to considerable heights above the earth's surface, where *Clouds* are formed; the small water droplets grow in size, and finally fall to the ground. There is no clear distinction between rain and *Drizzle*. See *Rainfall*.

RAINBOW. The coloured bow consisting of the colours of the spectrum, seen when sunlight falls on rain, being caused by the reflection and refraction of the light in the water drops. The red is on the outside edge of the bow, the violet on the inside, in the *primary bow*. Some of the light falling on to the water drops, however, is twice reflected, and a *secondary* or *outer bow* is formed, with the colours in the reverse order to those of the primary bow, and much less distinct. The primary bow has an angular radius of about 42°, the secondary bow one of about 54°.

RAINFALL. The total amount of rain deposited on a given area during a given time, as measured by a *Rain Gauge*. Melted snow and hail are included with the rain, and this *Precipitation* from the clouds constitutes most of the rainfall; very small additions are made for dew, hoar frost, and rime. Three different types of rainfall are generally recognized, depending on the process by which the clouds were formed: *Orographic*, *Cyclonic*, and *Convectional*. The comparatively heavy rainfall received on the mountains of western Scotland is largely orographic; cyclonic rainfall, due to depressions, provides much of the rain of the eastern half of the British Isles and continental Europe; convectional rain is typical of the equatorial regions. The rainfall of a place is very largely dependent upon the prevailing winds; it varies seasonally in many regions according to the movement of the sun and the accompanying movement of the rain belts. For climatological purposes, the most valuable rainfall data are the mean deposits for different areas for the different months or seasons, or for the whole year.

RAIN GAUGE. An instrument which is used for measuring *Rainfall*. It consists of a funnel, usually of diameter 5 in. or 8 in. at the mouth, fitted into a glass vessel, which holds the rain that has fallen on to the funnel. The water cannot evaporate from this vessel, and when it has been collected for a certain period, it is measured in another vessel graduated according to the area of the mouth of the funnel. The rainfall is thus expressed as a depth, usually in millimetres or inches, and represents the depth to which the ground would have been covered with rain, assuming that none could escape, since the rain gauge was last read. The rain gauge is normally read twice a day. See *Hyetograph*.

RAIN SHADOW. An area which has a relatively light average rainfall because it is sheltered from the prevailing rain-bearing winds by a range of mountains or hills; it is therefore situated on the lee side of the range. On the windward slopes, by contrast, the rainfall is extremely heavy, owing to the forced ascent of the moisture-laden air. This air is warmed and dried by descent on the lee side, so that little rain is deposited there. One of the best-known examples is provided by the Western Ghats in India: the western slopes, exposed for about five months to the SW. monsoon, have a very heavy rainfall, in places exceeding 200 in., but on the eastern slopes – in the rain shadow – the rainfall at many places is 25 in. or less.

RAIN SPELL. In the British Isles, a period of at least fifteen consecutive days, each of which has had at least .01 in. of rain. See also *Absolute Drought*.

RAISED BEACH. A beach which has been raised by earth movement to form a narrow *Coastal Plain*; it is often bounded by inland cliffs. If more than one rise has taken place, there may be raised beaches at different levels.

RAND (Afrikaans). In South Africa, a low ridge of hills, often covered with scrub. The term 'The Rand' is often used specifically for the Witwatersrand, in the southern Transvaal, famous for its rich gold-bearing reefs.

RANGE. A line of mountain ridges, with or without peaks, in which the crests are relatively narrow; if the altitude is comparatively low, the ridges may constitute a range of hills rather than of mountains.

RANGE OF TEMPERATURE. The difference between the highest and lowest temperatures of a place during a certain period. For climatological comparisons, the value most often used is the *mean annual range* of temperature, the difference between the mean temperatures of the warmest and coldest months. Mean annual ranges are very high in the midst of the great land masses, e.g. in Siberia, but relatively low in the southern hemisphere, owing to the moderating influence of the oceans. At Verkhoyansk, Siberia, for instance, the mean annual range is 66° C.; at Quito, Ecuador, practically on the equator, the mean annual range is ·4° C., though the diurnal range is about 20° C. The *annual extreme range* is the difference between the mean extremes of the warmest and coldest months. The *absolute range* of temperature is the difference between the highest and lowest temperatures ever experienced at a place.

The *mean diurnal range* for any period is the mean difference between the highest and lowest temperatures for each day of that period, a month being usually chosen, over a number of years. Its variation in the different regions is less than that of the mean annual range; both ranges, however, depend very largely on the geographical position of the place. In many regions, especially near the equator, the mean diurnal range is much greater than the mean annual range: for, to obtain the latter, the mean monthly temperatures are first calculated, and this causes the daily maximum and daily minimum temperatures partially to neutralize each other.

RAPIDS. Part of a river where the current is flowing with more than normal swiftness. They may be caused by a sudden steepening of the slope, or by unequal resistance in the successive rocks traversed by the river. They are often due to *Outcrops* of an unusually hard rock. At the outcrops, erosion is less than on the softer rock, and the slope of the river above each outcrop thus decreases, while below it the slope increases; over the hard rocks, then, rapids or *Waterfalls* are formed. If the layer of hard rock dips gently downstream, rapids rather than waterfalls are generally produced.

RASPUTITSA. In Siberia, the period of several weeks between the winter, when the ground is snow-covered, lakes and rivers are frozen, and transport is largely by sledge, and summer, when the ground is again dry, and transport is by wheeled vehicles over the now usable roads. It is a period of extensive floods and mud, owing to the melting of snow and ice, and transport of any kind is extremely difficult.

RATIONAL HORIZON. See *Horizon, Rational or True*.

RAVINE. A long, narrow depression in the earth's surface, rather smaller than a *Valley* but larger than a *Gully*. Several gullies often lead to a ravine, and several ravines to a valley.

RAW MATERIALS. Goods which are utilized by certain industries for manufacture into consumable commodities; they include partially manufactured goods, such as wood pulp and wheaten flour, which are afterwards subjected to further manufacturing processes. Some raw materials may become finished products, or vice versa: e.g. coal tar, formerly the finished product of the distillation process, is now the raw material of a host of chemical products.

RED CLAY. A type of *Ooze* which covers most of the deeper parts, the *Abyssal* region, of the ocean bed, at depths greater than about 2,000 fathoms, and consists mainly of inorganic material. It is composed very largely of minute particles of aluminium silicate, and is thus a true *Clay*, and it also contains volcanic material, especially *Pumice*, meteoric dust, and occasionally the teeth of sharks, etc.; its uniformly red colour is due to the presence of iron oxides. Red clay covers a very extensive area in the Pacific Ocean, a more restricted area in the Indian Ocean, and several parts of the Atlantic Ocean.

RED MUD. A marine deposit very similar to *Blue Mud*, and deriving its reddish colour from the presence of iron oxide. It consists of wind-borne *terrigenous* dust; it occurs off the tropical Atlantic coast of South America, and in the China Sea, where it is due to *Loess*.

REEF. (1) A ridge of rocks, lying near the surface of the sea, which may be visible at low tide, but is usually covered by the water. The most common type is the *Coral Reef*.

(2) A mineral vein, especially of gold-bearing quartz.

REFORESTATION. The planting of trees on land where a forest has previously stood, but has been destroyed, e.g. by a forest fire. See *Afforestation*.

REG. In the Sahara, the extensive areas of flat, desert plain from which the fine sand has been blown away, leaving the surface covered with small stones and gravel. See *Hammada*.

REGELATION. The process by which ice is frozen again, after it has been melted under pressure, or by which pieces of ice are joined together under pressure. The melting is due to the fact that the melting point of ice is lowered by pressure, so that on the release of that pressure the melting point rises, and the water freezes again. Regelation is believed to explain the movement of a *Glacier*.

REGOLITH. See *Mantle Rock*.

REJUVENATED RIVER. An old river which has been made young

again, e.g. by the uplift of the land over which it flows, so that it begins to cut into its channel once more.

RELATIVE HUMIDITY. The ratio between the actual amount of water vapour in a given volume of the air and the amount which would be present if the air were saturated at the same temperature, usually expressed as a percentage. It thus affords a measure of the relative dampness of the atmosphere, and is determined by means of a *Hygrometer*.

RELICT MOUNTAINS. Mountains which owe their present form to denudation, having been worn down, for instance, from a *Plateau* or a *Horst*, or from a chain of *Folded Mountains*, or simply from inclined strata. See *Circumdenudation, Mountains of; Monadnock; Inselberg*.

RELIEF. The differences in elevation of any part of the earth's surface.

RELIEF MAP. A map which seeks to depict the surface relief of an area. A *Contour* map and a *Photo Relief* map both achieve this object. A relief model shows surface relief in three dimensions, though not necessarily to scale. The vertical scale usually has to be much exaggerated; an inch on the vertical scale, for instance, may represent only one-tenth or even one-twentieth of the distance that it represents on the horizontal scale.

REVERSED FAULT. See *Thrust Fault*.

RHUMB-LINE or LOXODROME. A line on the earth's surface which cuts all meridians at the same angle. A vessel which sails along such a line is said to take a rhumb-line course; if proceeding from one point to another directly to the NE., for example, it will maintain a north-easterly direction throughout the voyage. If the two points are near the equator, or almost on the same meridian, the difference between the rhumb-line route and the *Great Circle Route* is comparatively small, but in high latitudes, when one point lies on a meridian far distant from that of the other, the rhumb-line route is considerably longer than the Great Circle route. On the map according to *Mercator's Projection*, the rhumb-line is depicted as a straight line.

RIA. A long, narrow bay or inlet into the sea coast, caused by the subsidence of the earth's surface in a region of ridges and furrows where these are not parallel to the coast, the valleys being thus drowned. It often branches at the end, and, unlike the *Fiord*, gradually deepens towards the sea; rivers which were once tributary to the main river valley often flow separately into the ria, and the latter is enclosed by the walls of the former valley. The most resistant rocks of the valley form islets in the ria. Some of the best-known rias are those of the coasts of SW. Ireland and NW. Spain. (See Fig. p. 148.)

RIDGE OF HIGH PRESSURE. An elongated area of high atmospheric pressure, forming an extension to an *Anticyclone* or high-pressure area, in the same way as a ridge projecting from a mountain. It is the opposite of a *Trough of Low Pressure.*

Rias along the coast of SW. Ireland.

RIFT VALLEY. A valley which has been formed by the sinking of land between two roughly parallel *Faults;* such a valley is long in proportion to its width. The best-known example is the rift valley which extends through Syria, Israel and Jordan, and East Africa, over 3,000 miles in length; it includes the Sea of Galilee, the valley of the River Jordan, and the Dead Sea, the Gulf of Akaba, and the Red Sea, and runs through Lake Rudolf and several smaller lakes to Lake Nyasa, with a branch through Lakes Tanganyika, Edward, and Albert. Examples of smaller rift valleys are the Rhine valley between the Vosges Mountains and the Black Forest, and the Central Lowlands of Scotland. Rift valleys are frequently known as graben (German) or trough faults.

RIMAYE. See *Bergschrund.*

RIME. A deposit of ice crystals formed when supercooled water drops come into contact with exposed objects and freeze on them. Rime that is formed in this way from the water drops in fog is usually known as *soft rime;* when it is formed by the freezing of water drops in *Drizzle,* it is usually known as *hard rime.*

RIVER. A stream of fresh water which for at least part of the year is larger than a brook or creek, and flows by a natural channel, being confined within banks, into the sea, or into a lake, or into another river. In a semi-arid region, however, a river may become no more than a series of pools during the dry season; in an arid region it may disappear in a desert. A river may originate, at the point known as its *source*, in a number of rivulets, a spring, a lake, etc.; the path it follows is known as its *course*; it gradually swells as it is joined by *Tributaries*, all the rivers together forming a *river system*; it enters the sea or a lake by its *mouth*. The *upper* parts of a river are situated near its source, the *lower* parts near its mouth.

RIVER BASIN. See *Basin* (2).

RIVER CAPTURE, RIVER PIRACY, or BEHEADING. The action of a river in acquiring the headstreams of a second river by enlarging its drainage area at the expense of the other. This process is carried out by the more powerful river, which erodes its valley more deeply and cuts back into the valley of its weaker neighbour. The headstreams of the latter are thus diverted into the former, and the latter is said to have undergone *beheading*, or have been *beheaded*; the bend at which the capture took place is known as the *elbow of capture*. See *Misfit River*.

RIVER PIRACY. See *River Capture*.

RIVER PORT. See *Port*.

RIVER PROFILE. See *Profile, River*.

RIVER TERRACE. A platform of land formed beside a river flowing across a plain, when for some reason it has commenced vertical *Corrasion* again. Its channel is thus deepened, and the *Flood-Plain* becomes a flat alluvial area well above the level of the river even at flood-time. After vertical corrasion has ceased, lateral corrasion will recommence, and a new flood-plain will be formed below the first one. If vertical corrasion again recommences, the second flood-plain, too, will form a river terrace, and the process may continue till a series of terraces at different levels has been created by the side of the river.

RIVULET. A small river.

ROADSTEAD. An open anchorage for ships, which may be sufficiently sheltered to give protection from heavy seas, usually by reefs, sandbanks, or islands.

ROARING FORTIES. The prevailing westerly winds which blow throughout the year over the oceans of the southern hemisphere, between about 40° and 60° S., and, being unobstructed by land masses, are regular and extremely strong. See *Brave West Winds*. The name Roaring Forties is also applied to the region where these winds blow, a region where *Depressions* continually follow one another all through the year, and the weather is stormy, rainy, and comparatively mild.

ROCHES MOUTONNÉES (French). The hillocks of rock of a glaciated valley (a valley down which a glacier has flowed), which have been smoothed and marked with *Striae* by the glacier on the upstream side, and left rough and rugged, with steeper slopes, on the downstream side. The upstream side, against which the glacier has flowed, is ground down by the stones embedded in the ice, but on the downstream side the glacier pulls away any blocks which are sufficiently loosened by the development of joints. Thus the appearance of the valley is often very different according to the direction from which it is viewed: looking down the valley, the rocks appear smooth and convex, while looking up the valley they appear rugged.

ROCK. One of the solid materials of which the earth's crust is mainly composed, being made up of *Minerals*; in some cases it consists of only a single mineral, but usually of several minerals. A rock, however, does not normally have a definite chemical composition, as a mineral does. The many different kinds of rocks in the earth's crust are divided into three major classes—*Sedimentary*, *Igneous*, and *Metamorphic*. More popularly, a rock is any large mass of the harder portions of the earth's crust.

ROCK FLOUR. The finely-ground solid matter produced by abrasion of the bed of a glacier by the stones and rocks embedded in its base, as they move forward with the glacier down the valley.

ROCK-GLACIER. A tongue or 'stream' of rock fragments, resembling a *Glacier* in form, and composed of *Scree*, which moves gradually downwards through the action of alternate frost and thaw and gravity; such rock-glaciers are especially numerous in Alaska.

ROPY or CORDED LAVA. A mass of *Lava* which has solidified so that its surface is smooth and exhibits ropy or corded shapes; the Hawaiian term pahoehoe is sometimes used.

ROTATION OF CROPS. The system of farming by which different crops are repeatedly grown on the same land in a specific order. For field crops the duration of a complete cycle usually varies between two years and eight years, and depends on the number of crops and their frequency of change. Among the advantages of a rotation are the more effective control of weeds, pests, and diseases, and the more economical utilization of soil food. The well-known Norfolk rotation, the first type to be generally adopted in England, is completed in four years: roots (e.g. turnips) are followed by barley, then grass and clover (which replenishes the soil with nitrogen), then wheat (which requires abundant nitrogen), in successive years.

RUN-OFF. That portion of the rainfall which ultimately reaches the streams; it consists of the water which flows off the surface, instead of sinking into the ground, together with some of the water which originally sank into the ground and joins it later in the streams. The part of the rainfall which leaves the ground surface at once and enters streams is called the *immediate run-off*. Most of the water that sinks into the ground eventually returns to the surface by seepage and from springs, and is called *delayed run-off*. The run-off is faster and greater (1) during heavy rain than during a protracted drizzle, (2) on clay soils than on sandy soils, (3) on frozen soils than on frostless soils, (4) in treeless areas than in the forests. The ratio between run-off and rainfall varies considerably with climatic conditions.

S

SADDLE-REEF. A lens-shaped strip of rock formed in a similar way to a *Phacolith*, but consisting of *Quartz*. The term is used principally by miners, and is sometimes wrongly assumed to be synonymous with phacolith.

SAGEBRUSH. A type of vegetation, common in the semi-desert regions of the western United States, where the common sagebrush predominates. Nevada, which has considerable areas of such vegetation, is sometimes known as the Sagebrush State.

ST ELMO'S FIRE. A luminous discharge of electricity sometimes seen during stormy weather at the extremities of tall objects, such as the mastheads and yard-arms of ships and the tops of trees.

ST LUKE'S SUMMER. A period of fine weather supposed to occur about St Luke's Day, October 18.

ST MARTIN'S SUMMER. A period of fine weather supposed to occur about St Martin's Day, November 11.

SALINA or SALINE. See *Salt-Pan*.

SALINITY. The degree of saltness of the oceans and seas, lakes and rivers, usually expressed as the number of parts per thousand, or the weight of salt dissolved in 1,000 parts of water. The mean salinity of sea-water, for instance, is 35 per thousand, but the salinity varies from over 40 per thousand in the Red Sea to about 30 per thousand in the polar seas. Near the equator there is a region of relatively low salinity, owing to the heavy rains. Bordering this belt are areas of higher salinity, for here the *Trade Winds* generally are dry, cause rapid evaporation, and the surface waters thus become salt. Beyond these areas, the salinity decreases polewards, for rainfall is greater and evaporation slighter.

Although river water contains a small quantity of mineral salts in solution, it is comparatively fresh. A partially enclosed sea like the Baltic Sea, which receives many large rivers, and also has heavy falls of rain and snow and small evaporation, thus has a salinity below 12 per thousand. The Black Sea, too, is comparatively fresh, with a salinity of about 17 per thousand. The Red Sea, on the other hand, receives no large rivers, has great evaporation and low rainfall, and thus has a salinity of over 40 per thousand. For similar reasons, the salinity of the Mediterranean Sea is over 37 per thousand. Seas and lakes with no outlet to the ocean have a still higher salinity, for the salts brought down by rivers have accumulated since their formation. The salinity of the Dead Sea is about 250 per thousand. See *Salt Lake*.

Of the 35 parts of dissolved salt contained on the average in 1,000 parts of sea-water, over 27 parts consist of sodium chloride or common salt, the other main constituents being magnesium chloride, magnesium sulphate, and calcium sulphate.

SALPAUSSELKA. In Finland, a ridge of sand and gravel running from east to west, usually interpreted as being a series of terminal *Moraines* formed in the quaternary *Ice Age*.

SALT DOME or SALT PLUG. A mass of salt which has been forced upwards by subterranean pressures, penetrating rock strata which originally lay above it.

SALT LAKE. A lake, situated in an arid region, which has no outlet and so accumulates salts brought down by the rivers that enter it. Loss of water by evaporation at least equals the supply of water by precipitation and inflow, and the lake gradually becomes more salty. The Dead Sea has a *Salinity* about 250 per thousand, the Great Salt Lake, Utah, U.S.A., about 220 per thousand, and part of the Caspian Sea about 170 per thousand. If its climate becomes less arid, a salt lake may become a fresh one in time; if the climate of a fresh lake becomes more arid, it may eventually become a salt lake.

SALT MARSH. A *Marsh* which at times is flooded by the sea, or an inland marsh in an arid region in which the water contains a high proportion of salt. The terms *salina* and *saline* are sometimes used to denote a salt marsh.

SALT PAN. A hollow, formerly containing water, in which a deposit of salt is left behind owing to evaporation of the water. The terms *salina* and *saline* are sometimes used to denote a salt pan.

SALT PLUG. See *Salt Dome*.

SAMOON. In Persia, a hot, dry, *Föhn* type of wind which blows from the mountains of Kurdistan. The term, alternatively spelt samun, is also sometimes used to signify the *Simoom*.

SAND. A mass of minute particles of mineral *Detritus* which is finer than *Gravel* but coarser than *Silt*. It is also not so fine as *Dust*, and is not normally lifted by the winds far above the earth's surface. Its movement is thus controlled by obstacles in its path, and it is often heaped up in mounds called *Dunes*. Most of the rock-making minerals of the earth's crust occur in sands, but comparatively few are common; of these, *Quartz* is by far the most abundant. When consolidated, sand forms a *Sandstone*.

SANDBANK. A submerged ridge of sand in the sea or a river often exposed at low water.

SAND-BAR. See *Bar* (2).

SANDR. See *Outwash Plain*.

SANDSTONE. A porous sedimentary rock consisting of grains of sand bound together by such substances as calcium carbonate or silica. Sandstones are characterized by the nature of the binding material, e.g. calcareous sandstone, and their colour largely depends upon it. Some sandstones are well cemented, containing considerable *Quartz*, and form extremely durable rocks; other sandstones are poorly cemented, and readily disintegrate on exposure.

SANDSTORM. A storm in which a mass of relatively coarse sand is blown along through the air by a strong wind. The sand particles are rarely raised more than 50 to 100 ft above the surface, and are not carried far from their source. A sandstorm should be distinguished from a *Duststorm*.

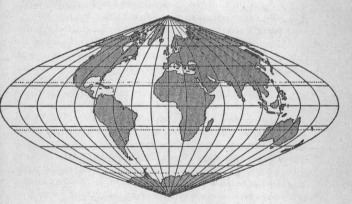

Sanson-Flamsteed Sinusoidal Projection.

SANSON-FLAMSTEED SINUSOIDAL PROJECTION. A modified *Conical Projection*, being a particular case of *Bonne's Projection*. The equator is chosen for the standard parallel, and thus is a straight line truly divided. The central meridian is a straight line at right-angles to the equator, and is truly divided; the parallels are horizontal straight lines at equal distances apart, and are truly divided, and the meridians on each side of the central meridian are formed by drawing curves through corresponding points on each parallel. The projection is termed sinusoidal because these meridians are sine curves. It is homolographic or equal-area, but shapes are distorted towards the margins; in atlases it is frequently used for the map of Africa, for the extent of this continent in latitude is divided nearly equally by the equator, and its extent in longitude is only moderate.

SANTA ANNA or **SANTA ANA.** The northerly to easterly wind of southern California which, blowing from the deserts and having been heated by descent from the mountains, is hot, dry, and dusty. Like the *Sirocco* of the Mediterranean area, it causes much discomfort to the inhabitants and considerable damage to crops by drying up the vegetation. It is mainly experienced in winter, being due to the passage of depressions, but it causes the worst destruction if it occurs in spring, when the fruit trees are in blossom, or when the young fruit is formed.

SAPROPHYTE. A plant which lives on decaying organic matter. Most of the saprophytes are fungi.

SASTRUGI. Ridges on the surface of the snow, caused by winds and varying in size according to the strength and duration of the wind and the condition of the snow surface: mainly applied to the U.S.S.R., but sometimes to snow-covered areas elsewhere.

SAVANNA or **TROPICAL GRASSLAND.** The region which borders the equatorial forest in each hemisphere, and thus lies between the latter and the hot deserts; the natural vegetation is mainly grass with scattered trees, for there are distinct wet and dry seasons, and the lack of rainfall during the latter prevents the growth of forests except in particularly moist places. The rains occur during the hot season, when the belt of equatorial calms moves to the savanna, but in the opposite season the savanna is under the influence of the drying *Trade Winds*. Thus in the hot season the vegetation freshens and grows vigorously, while in the opposite season the ground is parched and vegetation withers. The *Llanos* and *Campos* of South America are particular examples of savannas, but the most extensive savanna land is in Africa; the climate, in fact, is often known as the Sudan type. Because of the natural covering of grass, great herds of cattle are reared on the savanna.

SCARP.* See *Escarpment.*

SCHATTENSEITE (German). See *Ubac.*

SCHIST. A metamorphic rock which has been rendered cleavable into many thin plates through the effect of intense heat and pressure. Schists are named by their most conspicuous mineral, e.g. mica schist. Nearly all types of *Sedimentary* and *Igneous*

Rocks become schists if sufficient heat and pressure are applied.

SCORIA. A volcanic *Lava* which is extremely vesicular owing to the escape of gases while it was still viscous; it is formed during a violent eruption, and may be looked upon as a very coarse type of *Pumice*.

SCOTCH MIST. A type of precipitation consisting of a mass of minute water drops, resembling both *Mist* and *Drizzle*; it occurs usually in hilly or mountainous areas, and is due to the presence of dense clouds near to the ground. The name is derived from the fact that the phenomenon is common in Scotland, owing to the mountainous character of most of the country; elsewhere it is sometimes known as 'mizzle' – a combination of 'mist' and 'drizzle'.

SCREE. See *Talus*.

SCRUB. A dense mass of low-growing evergreen plants, about four to six feet high, with occasional taller trees. It is thus usually found in regions which have insufficient rainfall or too poor a soil for forest growth, e.g. in semi-arid areas on the margins of the hot deserts; the inhabitants are generally forced to follow a nomadic, pastoral life.

SCUD. Ragged fragments of very low cloud driven along by a strong wind, often beneath ordinary rain clouds.

SEA. One of the smaller divisions of the oceans, especially if partially enclosed by land, e.g. Mediterranean Sea, North Sea; a large expanse of inland salt water, even if completely land-locked, e.g. Caspian Sea. The name is also loosely applied to the great mass of salt water which covers much of the earth's surface.

SEA BREEZE. The diurnal movement of air from the sea to the land, caused by differential heating. During the day, the greater heating of the land causes the air to ascend, and air from the sea moves in to take its place. The sea breeze, like the *Land Breeze*, is thus most noticeable and most regular when temperature changes are most regular, i.e. when the *Pressure Gradient* is slight and the sky is clear; it then commences during the morning, and continues till the early evening, reaching maximum strength in the afternoon. It varies considerably in strength along the same coast, being influenced by topography. In temperate regions it does not usually exceed about 10 miles per hour, though in the tropics it may reach 20 to 25 miles per hour; it normally extends a few miles on each side of the coast-line. Like the land breeze, it reaches its greatest development in the *Doldrums*, where the pressure gradient is slight; here it does not, as elsewhere, merely modify the prevailing winds.

SEA LEVEL. (1) The level which the surface of the sea would assume if uninfluenced by tides, or waves, or swell.

(2) The mean level between high and low tide at any place. See *Mean Sea Level*.

SEA-MILE. A *Nautical Mile.*

SEAPORT. See *Port.*

SEASONS. Those periods of the year which are characterized by special
climatic conditions, mainly caused by the inclination of the
earth's *Axis* to the plane of the *Ecliptic* and the revolution of the
earth about the sun. In the temperate regions, four seasons, each of
three months' duration, are generally recognized: in the northern
hemisphere spring includes March, April, and May: summer, June,
July, and August: autumn, September, October, and November,
and winter, December, January, and February; in the southern
hemisphere the seasons are the opposite of these, with spring including September, October, and November, and so on. In areas
where the oceanic influence is considerable, as in Great Britain, the
seasonal variations are not nearly so large as in the interior of the continents, and the change from season to season is much more gradual.
In regions outside the temperate zone, the year cannot be divided
into four seasons. In much of the tropics, for instance, the seasons
depend more on rainfall than on the position of the sun, and the
year is usually divided into the wet or rainy season and the dry
season; over much of India, and elsewhere in the monsoon regions,
the year is divided into cold season, hot season, and rainy season. In
the polar regions the changes between winter and summer take
place so abruptly that spring and autumn are scarcely recognizable.

SECOND. (1) A unit of measurement of time equal to one-sixtieth of a
minute.

(2) A unit of measurement of latitude or longitude equal to one-
sixtieth of a *Minute.*

SECONDARY DEPRESSION. A comparatively small area of low atmospheric pressure associated with a main or primary *Depression.* It
accompanies the primary depression, frequently revolves round the
latter as they both move along their course, and in time may
dominate the system and absorb the primary depression.

SECULAR. Extending over a very long period of time.

SEDIMENTARY ROCKS. Rocks which have been deposited as beds,
often as sediments (i.e. under water), forming one of the three main
types of rocks which make up the earth's crust. They were laid
down in distinct layers or strata, separated by *Bedding Planes*, principally by the sea, but also by lakes, or streams, or glaciers, or even
by the wind; they are therefore often known as stratified rocks.
Formerly they were sediments similar to the sands, gravels, and
muds now being deposited in the seas, lakes, and rivers. The strata,
which vary in thickness from a few inches to several feet, are sometimes horizontal, sometimes tilted; a considerable tilting of the
strata indicates that the materials were disturbed after they had

been deposited. Among the common forms of sedimentary rocks are *Sandstone*, *Shale*, *Limestone*, and *Conglomerate*, while coal is a sedimentary rock of organic origin.

SEEPAGE. The slow oozing out of *Ground Water* on to the earth's surface, as distinct from the more pronounced flow of a *Spring*.

SEICHE. A variation in the level of the surface of a lake, somewhat resembling a tide, first observed on the Lake of Geneva; the rise and fall of the water may vary between a few inches and several feet. The causes of seiches include changes in atmospheric pressure, small earthquakes below the bed of the lake, and the wind. Seiches usually take place in the direction of the longest diameter of a lake, but are occasionally transverse.

SEIF DUNE. A type of sand *Dune*, common in the Sahara, which consists of a long sharp ridge lying parallel to the direction of the prevailing wind. The latter tends to increase the length of the dune, while cross-winds tend to increase its height and width.

SEISMIC FOCUS. The place below the earth's surface where an *Earthquake* originates, and from which the vibrations spread in all directions; it is usually several miles beneath the surface. It is now held that an earthquake does not originate at a point, but along a line.

SEISMOGRAPH. An instrument for recording *Earthquake* shocks. The principle of its action consists in the disturbance of a portion of the apparatus by earth tremors, the suitable amplification of the motion thus produced, and then its recording. The vibrations used to be recorded by a pen tracing on a revolving drum, but in more modern instruments a photographic record is obtained on a moving film. Many seismographs are so sensitive that they will record vibrations due to an earthquake thousands of miles away, and its distance and also its direction may be approximately calculated. A properly equipped seismological observatory has three different seismographs; two of these will record the horizontal components of the earth's vibrations, one in an east-west and the other in a north-south direction, while the third will record the vertical components.

SEISMOLOGY. The science of the study of *Earthquakes*.

SEISTAN. A strong northerly wind, sometimes exceeding 70 miles per hour, experienced throughout the summer in the province of Seistan, in eastern Persia. It is known as the 'wind of 120 days'.

SELVAS. The dense *Equatorial Forest* region of the basin of the River Amazon in South America.

SENSIBLE HORIZON. See *Horizon, Sensible, or Visible*.

SÉRACS (French).* Pinnacles or pillars of ice of various shapes into which a glacier is sometimes broken up when it reaches a steep slope, numbers of *Crevasses* being formed in all directions and crossing one another.

SERICULTURE. The culture of silkworms for the production of raw silk.

SEXTANT. An instrument used for measuring the angular distance between two objects. Very similar in design to the *Quadrant*, it represents an improved form of the latter. It is most widely used in navigation, for by its means the apparent altitudes of celestial objects may be measured, so that the position of the observer in degrees of latitude and longitude may be calculated.

SHAKE-HOLE. A term used in the limestone districts of the Pennines, in England, for a *Sink Hole*.

SHALE. A finely-grained *Sedimentary Rock* produced from *Clay*; it may have been formed by compression due to the weight of overlying rocks, or by cementation, e.g. by lime. Shales are rarely pure, and they are characterized by the main foreign substance they contain or by the type of cement. They are usually finely stratified, and thus readily split into thin layers, and they are so soft that they readily disintegrate.

SHAMAL. The extremely constant north-westerly wind, experienced chiefly during the summer in Iraq, which blows across the Tigris-Euphrates plains, being controlled – like the *Etesian Winds* of the eastern Mediterranean and the *Seistan* of eastern Iran – by the seasonal summer low over north-west India and Baluchistan. Being uninterrupted by depressions, it blows with great regularity. It is particularly strong during the day, when it carries clouds of dust, and causes severe duststorms, especially in southern Iraq, but there is normally a marked lull at night. The shamal is also experienced in winter, but with far less regularity.

SHEEP-TRACKS. See *Terracettes*.

SHINGLE. A collection of loose pebbles often found on the seashore.

SHIP CANAL. A *Canal* constructed of more than usual depth and width to permit the passage of ocean-going vessels. Its function is usually to shorten the voyage between two seas by cutting through an *Isthmus*. The most important examples of this type are the Suez Canal, joining the Atlantic Ocean through the Mediterranean Sea with the Indian Ocean, and the Panama Canal, joining the Atlantic Ocean to the Pacific Ocean. Sometimes a ship canal is constructed from the sea inland, in order to convert an inland place into a seaport; an example is the Manchester Ship Canal.

SHOAL. (1) A ridge of sand or of rocks just below the surface of the sea or of a river, and therefore dangerous to navigation.

(2) A multitude of fish swimming near to the surface of the sea.

SHOOTING STAR. See *Meteor*. Practically the only relation that a shooting star bears to a true star is that, being seen at night against the sky, it *appears* to move among the stars.

SHORE-LINE. See *Coastline*.

SHOTT. A shallow, temporary *Salt Lake* or *Salt Marsh* found on the Algerian plateau and in the valleys south of the Atlas Mountains. The shotts extend for about 250 miles over the plateau, which is often known as the Plateau of the Shotts. The dry basins left behind when the lakes have disappeared are characterized by deposits of salt and often by the absence of vegetation.

SIAL. The term used to denote the comparatively light rocks of the *Lithosphere*, including granite, lying beneath the continents. The name is derived from the initial letters of silicon and aluminium, the principal elements contained in these rocks. See *Sima*.

SIDEREAL DAY. The period of time during which a star describes a complete circle in its apparent journey round the *Pole Star*, representing the period of rotation of the earth on its axis, and equal to 23 hours 56 minutes 4 seconds. It is thus about 4 minutes shorter than the Mean Solar Day. See *Solar Day*, *Mean*.

SIERRA. An elongated range of mountains with a jagged crest, mainly in Spain and Spanish America.

SILL. A sheet of igneous rocks, often almost horizontal, formed when *Magma* forced its way between two layers of sedimentary rock, and there cooled and solidified; its thickness may be a few inches or several hundred feet, but its horizontal extent is always great in comparison with its thickness. A sill gives rise to *Dykes*; again, it often protects softer rocks beneath, and owing to unequal erosion may lead to the formation of a ridge or a flat-topped mountain. See *Laccolith*.

SILT. A deposit which is laid down in a river, lake, etc.; it is finer than *Sand* but coarser than *Clay*.

SIMA. The term used to denote the comparatively heavy rocks of the *Lithosphere*, including basalt, lying at a lower level than the *Sial* and believed to lie beneath the ocean basins. The name is derived from the initial letters of silicon and magnesium, the principal elements contained in these rocks.

SIMOOM. A hot, dry, suffocating wind or whirlwind experienced chiefly during spring and summer in the Sahara and the Arabian deserts. It usually carries with it dense masses of sand, which reduce visibility to a few yards or almost zero, and sometimes change the shape of the sand dunes along its track.

SINK HOLE or SWALLOW-HOLE. A saucer-shaped depression in the earth's surface, usually found in a limestone region, through which water may enter the ground and pass along an underground course. It is caused by the solvent action on the rock of rain water containing carbon dioxide from the atmosphere, and may be enlarged by the sinking of the ground. The term *Pot-hole* is sometimes applied to a sink hole.

SINTER. A mineral deposit from a *Hot Spring* or *Geyser*, which sometimes occurs in mounds, cones, or terraces. There are two principal kinds, calcareous sinter and siliceous sinter, depending on composition. The former is often known as *Travertine*, the latter as *Geyserite*.

SIROCCO or SCIROCCO. The southerly wind experienced in North Africa, Sicily, and southern Italy, which, blowing from the Sahara, is hot, dry, and sometimes dust-laden. Where it descends from a mountain range, as on the Algerian coast, its heat and aridity are increased. It often crosses the Mediterranean, picking up moisture, and reaches southern Italy as a hot, humid, and very enervating wind. The sirocco is formed ahead of one of the depressions which pass from west to east along the Mediterranean; it may therefore occur in all seasons, but it is very rare in summer, and is specially important in spring, when the depressions are most vigorous and the desert is already hot. It usually lasts for a day or two, and is then replaced by a much cooler northerly wind behind the depression. The sirocco withers vegetation and often causes much damage to crops, especially if it blows while the vines and olives are in blossom. In Egypt the wind is known as the *Khamsin*, in south-eastern Spain as the *Leveche*, in Tunis as the *Chili*, in Libya as the *Gibli*.

SKERRY. A rocky islet.

SLATE. A dense, fine-grained *Metamorphic Rock* produced by heat and pressure from a fine clay; it readily splits into thin, smooth plates suitable for roofing purposes. The true slates do not usually split along the bedding, but along planes of *Cleavage*.

SLEET. In Great Britain, precipitation consisting of a mixture of snow and rain. In the United States, the term signifies the ice pellets which consist either of frozen rain drops or of snow which has partly melted and has been re-frozen when falling through a cold layer of air near the earth's surface.

SLICKENSIDES. Rock surfaces on either side of a *Fault Plane* which have been polished or marked with *Striae* by friction between the moving blocks.

SMALL CIRCLE. A circle on the earth's surface which does not bisect the globe: e.g. any parallel of latitude except the equator. See *Great Circle*.

SMOG. A *Fog* that is heavily laden with smoke, and is therefore commonest in industrial and densely-populated urban areas. The term was coined from the two words 'smoke' and 'fog'.

SMOKES. The heavy morning and evening mists experienced during the dry season along the Guinea Coast of Africa, corresponding to the *Cacimbo*.

SNOW. *Precipitation* which takes the form of ice crystals of a delicate, feathery structure, having been formed from water vapour in the atmosphere at temperatures below the Freezing Point. The snow may fall in the form of individual crystals, or as large flakes produced by the amalgamation of a large number of crystals. Often the snow is melted during its descent, and reaches the surface as rain; hence on low-lying ground rain may be falling, while on neighbouring hilltops, where the temperature is lower, freshly-fallen snow may be observed. Ten inches or a foot of snow is regarded as equivalent to approximately one inch of rain, but the actual ratio varies considerably with the nature of the snow; for meteorological records the snow is melted, and its amount expressed as the equivalent depth of rain.

SNOWDRIFT. Snow which has been driven into a bank or heap by the wind, sometimes by accumulation in a sheltered spot, sometimes by transportation of fallen snow.

SNOWFIELD. A region of permanent snow; as temperatures must be low, it is common only in mountainous areas or in high latitudes.

SNOW-GAUGE. An instrument used for measuring the depth of snowfall. Most types are merely *Rain Gauges* which have been adapted to enable them to collect the snow, so that it can then be melted and measured. See *Snow*.

SNOWLINE.* The line on a mountain or hill slope which represents the lower limit of perpetual snow; below the line, any snow which falls is melted during the summer. It is not a sharply defined line for towards the margin of perpetual snow the covering becomes, patchy and irregular, until finally it disappears altogether. The altitude of the snowline varies considerably in different regions, and in general falls from the tropics to the polar regions. Near the equator it is about 16,000 to 20,000 ft above sea level, in the Alps and Pyrenees about 8,000 to 10,000 ft, in northern Scandinavia about 4,000 ft, while in the higher polar latitudes it is down to sea level. The altitude depends largely on the summer temperatures, which determine the rate of melting of the snow; exposure is an important factor, and the snowline is usually higher on the equatorial than on the polar side of a mountain. On a steep slope much of the snow descends as *Avalanches*, so that on a gentle slope, where most of the snow lies till it melts, the snowline is lower. But the most important influence on the snowline, along with summer temperature, is the amount of winter snow; other factors being equal, the snowline is higher in a dry region than in a wet region. In the Himalayas, for instance, the snowline is about 2,000 ft higher on the northern slopes than on the southern slopes, the higher temperatures of the latter being more than counterbalanced by the heavier precipitation due to the south-west monsoon.

SOIL. The loose material which forms the upper layer of the *Mantle*

Rock, consisting mainly of very small particles. It is penetrated by the roots of plants, which derive from it both food and moisture; in fact, from the standpoint of agriculture, soil constitutes only the few inches of the top layer of mantle rock in which cultivated plants are grown. The depth of the soil, however, actually varies from a fraction of an inch to many feet. Over 90 per cent of the dry substance of the average soil is inorganic mineral matter, produced from the original rock material by *Weathering*, but it also contains *Humus*. In the pore spaces of the soil proper, too, are liquid and gaseous components, in the form of water and air.

SOIL CREEP. The slow, almost imperceptible, but continual movement of the surface soil and rock fragments down slopes. When in dry weather the sun's heat causes the soil to crack, the crack opens downhill; when the soil fills with rain water, the crack is closed, again downhill. There is, then, a downward movement of the soil amounting to a fraction of an inch. A similar effect is produced by the expansion and contraction due to heat and cold. See *Solifluction*.

SOIL EROSION. The wearing away and loss of *Topsoil*, mainly by the action of wind and rain. *Sheet erosion* occurs, chiefly on sloping farmland, when the rain washes away a thin layer of topsoil. In *gully erosion* the rain forms small channels down the slope and these develop into gullies. *Wind erosion* is usually experienced in dry regions: the topsoil becomes loose and powdery, and the wind then carries it away.

SOIL GROUP, GREAT. An extensive group of soils having similar internal characteristics. The great soil groups include tundra soils, desert soils, *Podzol* soils, *Chernozem* soils.

SOIL HORIZON. See *Horizon, Soil*.

SOIL PROFILE. See *Profile, Soil*.

SOIL SERIES. A group of soils having similar profiles (see *Profile, Soil*), except for the texture of the surface soil, and developed from a particular type of parent material.

SOIL TYPE. A term used in the classification of soils, signifying the same as *Soil Series*, except that the texture of the surface soil must only vary within narrow limits. It is the principal unit used in soil mappings as well as other soil studies.

SOLANO. An easterly wind which often brings rain to south-eastern Spain and the Straits of Gibraltar. See *Levanter*.

SOLAR CONSTANT. The intensity of the sun's radiation in space at the mean distance of the earth from the sun, usually expressed in calories per square centimetre per minute.

SOLAR DAY, MEAN. The average length of the solar day, equal to 86,400 seconds – the 24 hours or day of civil time. Owing to the eccentricity of the earth's *Orbit* and the inclination of the equator to the *Ecliptic*, the solar day varies slightly in length at different times of the year, so that the average length, the mean solar day, is taken. The solar day is the interval of time between successive occasions when the sun is in the meridian of any fixed place; as the earth travels in its orbit round the sun in the same sense of rotation as that of its rotation about its axis, the solar day is slightly longer – nearly four minutes longer – than the *Sidereal Day*. To view the problem from another angle, the stars are at such a great distance from the earth that their apparent position is little affected by the earth's movement round the sun: the sun is so much nearer to the earth that its apparent position is materially affected, and it appears to lag behind each day. As the earth makes 1/365 of its revolution round the sun each day, the change of position due to one day's apparent annual motion of the sun equals the change of position due to about 1/365 of 24 hours, or nearly 4 minutes, of apparent daily motion.

SOLAR SYSTEM. The group of celestial bodies consisting of the *Sun*, the *Planets*, which revolve round it, the planetoids or *Asteroids*, *Comets*, *Meteors*, and *Meteorites*, and the satellites which revolve round the planets, e.g. the *Moon* round the earth. It does not include the stars, which are other suns at vast distances from the earth.

SOLFATARA. The vent associated with a *Volcano* which no longer ejects molten *Lava* or ashes, but continues to give off steam and certain gases; in distinction from the *Fumarole*, the exhalations from the solfatara consist principally of sulphuretted hydrogen and other sulphurous gases.

SOLIFLUCTION. Soil flow: the movement of soil and rock fragments down slopes, being more rapid than *Soil Creep*; it occurs mainly in the *Tundra*, where the top layer of soil remains saturated, and alternate freezing and thawing take place.

SOLSTICE. The time during summer or winter when the sun is vertically above the point which represents its farthest distance north or south of the equator, i.e. when the *Declination of the Sun* reaches its maximum or minimum: at this period the altitude of the sun at noon seems to be practically unchanged for a few days. In the northern hemisphere, the day on which the sun is vertical over the *Tropic of Cancer*, its farthest north, about June 21, is known as the *summer solstice*, and represents midsummer. The *winter solstice*, about December 22, is the day when the sun is vertical over the *Tropic of Capricorn*, its farthest south, and represents midwinter of the northern hemisphere. The days are longest and nights are shortest at the summer solstice, while the days are shortest and nights longest at the winter solstice. As the seasons in the southern hemisphere are the opposite of those of the northern hemisphere, the winter solstice is about June 21, and the summer solstice about December 22.

SONNENSEITE (German). See *Adret*.

SOROCHE. The type of *Mountain Sickness* suffered on the *Puna* of the Andes, sometimes affecting even the natives when they travel from the coast to the puna.

SOTCH. A term used in parts of France for a *Sink Hole*.

SOUND. A narrow passage of water or *Strait*.

SOUNDING. The process by which the depth of the sea is determined. By the simplest method, a leaden weight known as the sounding-lead is attached to a sounding-line, and is then lowered into the sea until it strikes the bottom; the length of line run out from the ship reveals the depth of the sea. By a more modern method, a sound wave is transmitted to the ocean bed, where it is reflected back to the surface, and the time interval between transmission of the wave and reception of the echo, combined with the known velocity of sound in water, enable the depth to be calculated. See *Fathometer*.

SOUNDING BALLOON. See *Ballon Sonde*.

SOURCE. See *River*.

SOUTHERLY BURSTER or **SOUTHERLY BUSTER.** The *Cold Wave* from a southerly direction experienced in southern and south-eastern Australia, usually behind a *Trough* of low pressure extending from the Antarctic. It is accompanied by typical *Line Squall* conditions, often being heralded by a long roll of cloud. The wind changes suddenly from a warm northerly, the *Brickfielder*, to the violent cold southerly, often laden with dust and generally accompanied by a thunderstorm. The fall of temperature is sudden and considerable, usually amounting to 10° C., often to much more. The Southerly Burster is most frequent in spring and summer, and is felt most severely along the coast of New South Wales; the mountains probably impede the advance of the following anticyclone, a steep *Pressure Gradient* is created, and the southerly wind blows with extreme violence. It corresponds to the *Pampero* of South America.

SOUTHERN CIRCUIT. The general southerly route by which depressions cross the North American continent from west to east, via the central United States. Towards winter, depressions show an increasing preference for this route over the more northerly routes, and by midwinter may be travelling as far south as the shores of the Gulf of Mexico. See *Northern Circuit*.

SOUTHERN CROSS. A constellation of four stars visible in the southern hemisphere, arranged in the form of a cross, the longer axis of which points almost due south; if this axis is extended to 4½ times its length, a point nearly over the South Pole is given. The constellation may thus be used in the southern hemisphere, like the

Pole Star in the northern hemisphere, for finding directions.

SOUTH MAGNETIC POLE. See *Magnetism, Terrestrial.*

SOUTH POLE. See *Poles.*

SPECTRE OF THE BROCKEN. See *Brocken, Spectre of the.*

SPELEOLOGY. The study of caves.

SPHERE OF INFLUENCE. A term formerly used to denote a region, usually defined by treaty, in which certain nations agreed that one or more of them should have exclusive liberty of action. It was also used rather loosely to denote any territory in which a foreign power sought to exert exclusive influence without annexation.

SPINNEY or SPINNY. A small wood with undergrowth.

SPIT. A narrow, low-lying tongue of sand or gravel, or small *Point*, projecting into the sea. It differs from a *Bar* in that it is attached to the land at one end, and it is formed by *Longshore Drift* across the entrance to a coastal inlet.

SPRING. A continuous or intermittent natural flow of water from the ground; it is formed when rain water sinks through the ground to a certain point, where it accumulates and finally gushes out. Its site thus depends on the position of the *Water Table*, the shape of the land surface, and the types of rock. A spring may be formed, for instance, when water percolates through a layer of porous rock till it reaches an impermeable layer, along which it runs till it gushes out at the surface; it may be formed when the land surface, e.g. in a valley, is below the level of the adjacent water table; it may be formed, too, when water, sinking through fissures in the ground, is led by them to the surface at some distant point below the level of intake. The water of a spring may be cold or warm, soft or hard. See *Geyser, Hot Spring, Mineral Spring.*

SPRING EQUINOX. See *Equinox.*

SPRING TIDES. The *Tides* of great amplitude when the earth, the sun, and the moon are practically in the same straight line; the gravitational pull of the sun acts in the same direction as that of the moon, and thus reinforces it. The high tide is higher and the low tide lower than usual. A consideration of the phases of the moon will show that they occur twice a month, about the time of new moon and full moon. See *Moon, Phases of.*

SQUALL. A violent wind which rises suddenly, has a brief duration (normally some minutes), and dies away suddenly; a temporary change of direction is often associated with it. See *Line Squall.*

STACK. A rocky islet or pillar, near to a coastline, which has been isolated by the erosive action of the waves.

STALACTITE.* A column of mineral matter hanging from an elevated point, in appearance much resembling an icicle; the commonest example is the column of calcium carbonate often found hanging from the roof of a limestone cave. This is produced from rain water which, with the help of carbon dioxide from the atmosphere, has dissolved the calcium carbonate in percolating through a bed of limestone or chalk, and then trickles down through cracks in the cave roof. As a drop of this water hangs, it partially evaporates, and leaves behind a small quantity of calcium carbonate; then the next drop leaves a further small quantity, and the deposit grows downwards, the water tending to trickle down its side and hang in drops at the lower end. See *Stalagmite*.

STALAGMITE.* A column of calcium carbonate formed on the floor of a cave by water which contains calcium carbonate in solution falling from the roof. Some of the water evaporates, leaving behind a solid deposit of calcium carbonate in the same way as with a *Stalactite*, and the stalagmite grows upwards. It is usually shorter and thicker than the stalactite. A stalagmite is often produced by the water dripping from a stalactite, and sometimes stalactite and stalagmite meet, forming a complete pillar from floor to roof of the cave. The deposit of calcium carbonate formed on the walls and floor of the cave from water trickling down is also often known as a stalagmite.

STANDARD TIME or ZONE TIME. Time which is referred to the mean time of a certain meridian, and is fixed for use over an extensive area. In general, the standard meridians are chosen to differ from the Greenwich meridian by multiples of 15° or 7½°, i.e. by an exact number of hours or half-hours, and the world is thus divided into a number of time zones. Each zone extends in a west-to-east direction over about 15° longitude, or approximately 7½° on each side of the standard meridian, and all places within the zone follow the same standard time. In western Europe, the Greenwich meridian is the standard meridian, and *Greenwich Mean Time* is the standard time; two other standard times are used in Europe, Central European Time and Eastern European Time. In countries which cover a considerable extent of longitude, several zones have to be utilized; in Canada and the United States, for example, there are Atlantic, Eastern, Central, Mountain, and Pacific Times, all differing from one another and from G.M.T. by multiples of one hour. The standard time system eliminates the confusion of having a different *Local Time* for all places situated on the different meridians round the globe, or a national civil time based on the local mean time of the principal national observatory.

STATUTE MILE. See *Mile, Statute*.

STEPPES. The mid-latitude *Grasslands*, consisting of level, generally treeless plains, of Eurasia, extending over the lower regions of the Danube, and in a broad belt over southern European U.S.S.R. and south-west Siberia. The term is sometimes applied to the corresponding mid-latitude grasslands in other continents; in North America, however, the commoner term is *Prairies*, in South America *Pampas*. The term is sometimes applied, too, to the semi-arid regions on the fringe of the hot deserts.

STEREOGRAPHIC PROJECTION. A *Zenithal* orthomorphic projection, in which the projection on to the tangent plane is made from the extremity of the diameter drawn at right-angles to the plane. Areas at a great distance from the centre are exaggerated, though much less so than in the *Gnomonic Projection*, and the shapes of large areas are somewhat distorted.

STEVENSON SCREEN. The standard shelter for thermometers on meteorological and climatological stations, to enable the true temperature of the air to be taken by protecting the thermometers from solar and terrestrial radiation. It consists of a wooden box, painted white and raised from the ground on a stand; the roof is double, with an intervening air space, the sides are louvred, one of them being hinged so as to serve as a door, and the air is thus permitted to enter freely, and the thermometers give a close approximation of the true air temperature. The ordinary pattern holds dry bulb and wet bulb thermometers and maximum and minimum thermometers, all of which are supported on a frame inside the screen as far as possible from roof, floor, and sides. A larger pattern also holds a *Thermograph* and *Hygrograph*.

STOCK. A term sometimes applied to a *Batholith* of relatively small size.

STONE AGE. The period when men used implements and weapons made of stone, preceding the period when they first extracted metals from their ores. See *Bronze Age*. It does not necessarily denote a fixed chronological period in history, for some peoples progressed beyond the Stone Age earlier than others.

STONE-RIVER. A 'stream' of boulders formed of *Scree* which gradually moves down a valley, being similar in form to a *Rock-Glacier*; such stone-rivers are well known in the Falkland Islands.

STOWED WINDS. Winds which are impeded by mountains, and are thus concentrated in gaps and on high plateaux, their normal velocity being thereby greatly increased.

STRAIT. A narrow stretch of sea connecting two extensive areas of sea. It may have been formed by fracture across an *Isthmus*, or by the sea overflowing land which had subsided, or by erosion.

STRATH. In Scotland, a broad river valley, in contrast to a *Glen*.

STRATIGRAPHICAL GEOLOGY or STRATIGRAPHY. That section of *Geology* which deals with the chronological succession of rock formations. The term is also sometimes used to signify the entire study of Historical Geology.

STRATOCUMULUS. A type of low *Cloud* consisting of a layer arranged in globular masses or rolls which are often so close together that their edges join; the layer is usually extensive.

STRATOSPHERE. The upper layer of the atmosphere, extending upwards from a height of about 6 miles above the earth's surface; the height at which it commences, however, is very variable, and is greatest at the equator. The base of the stratosphere also represents the layer at which the normal fall of air temperature with increasing height abruptly ceases. Within the stratosphere, the temperature is almost constant in the vertical direction, and the term 'isothermal layer' has sometimes been applied to it; this is misleading, however, as the temperature changes in a horizontal direction, and the ascents of *Sounding Balloons* have shown that the temperature actually increases with height in the lowest layers of the stratosphere. Temperatures are very low in the stratosphere, there are no clouds and practically no dust or water vapour, and no convection currents.

STRATUM. A more or less distinct layer of rock, occurring as one of a series of strata in the earth's crust. Rock strata vary in thickness from a fraction of an inch to several feet, and normally are horizontal, or nearly so.

STRATUS. A type of low *Cloud* consisting of a uniform layer, resembling fog which has been lifted from the ground.

STREAM. A course of running water; a rivulet, brook, or river.

STRIAE. Scratches or narrow grooves which have been worn on the surfaces of rocks when rock fragments have been dragged over them by a glacier.

STRIKE. The horizontal line along a rock stratum perpendicular to the direction of *Dip*. The relationship between strike and dip is thus the same as that between the contour and the slope of the ground.

SUBMERGED FOREST. A forest now covered by the sea, except possibly at very low tide, having been submerged by the sinking of the coastline. There are several examples round the coasts of the British Isles.

SUB-POLAR REGION. The region which is bordered on the poleward side by the tundra and on the equatorial side by the cool temperate or cool intermediate region or in some cases by the mid-latitude steppes and deserts. It is sometimes known as the cold temperate or cold intermediate region, and it practically coincides with the *Coniferous Forests* of high latitudes in the northern hemisphere. Winters are long and extremely cold, summers are very short: precipitation is relatively scanty, but on account of the slight evaporation is sufficient to promote forest growth.

SUBSEQUENT RIVER. A tributary to a *Consequent River*, flowing approximately in the direction of the *Strike*; it usually carves out a broad, deep valley because it flows over weak rocks, and becomes a powerful stream. It is so named because its formation is subsequent to that of the original *Consequent River*. See *Obsequent River*.

SUBSIDENCE. (1) The sinking of a portion of the earth's crust relatively to the surrounding parts.

(2) The slow descent of large masses of air towards the earth's surface, characteristic of anticyclones and of a quickly rising atmospheric pressure; the air is dynamically warmed and dried as it descends, so that subsidence is frequently associated with fine, dry weather.

SUBSISTENCE CROPS. See *Subsistence Farming*.

SUBSISTENCE FARMING or SUBSISTENCE AGRICULTURE. The type of farming in which the produce is consumed mainly by the farmer and his family and is not sold or traded. Crops so produced are known as *subsistence crops*. See *Cash Crops*.

SUBSOIL. The layer of rock particles lying below the true soil; it has less organic matter than the latter, and is apt to differ somewhat in its mineral content, being less exposed to oxidizing and hydrating agencies. It is less fertile than the true soil, but is penetrated to some depth by the roots of trees and plants.

SUB-TROPICAL REGION. The region lying between the tropics and the temperate or intermediate region. Unlike the tropics, it has a distinct seasonal rhythm, with a well-marked winter and summer, but the climate is generally warmer than in the temperate region. It is divided into the areas having a dry sub-tropical or *Mediterranean Climate* and those having a wet or humid sub-tropical climate – sometimes known as the *Cotton Belt Climate*; the former is experienced mainly on the western sides of continents, and is characterized by long, dry summers and mild, moderately rainy winters, while the latter is found on the eastern sides of continents, and has a higher rainfall, most of the precipitation falling, in general, during the summer.

SUDD. A floating mass of vegetable matter, applied particularly to that which collects on the White Nile, and at times has seriously impeded navigation; it consists principally of plants from the adjoining swamps. It sometimes forms a dam over 20 miles long and nearly 20 ft deep, and becomes so compact that it will support an elephant.

SUMATRA. A type of *Squall* experienced in the Malacca Strait, mainly at night and most frequently during the south-west monsoon. With the squall there is a sudden change of wind direction from southerly to westerly, and an abrupt fall of temperature; it is accompanied by a heavy bank of *Cumulonimbus* cloud, and generally by a violent thunderstorm and torrential rain.

SUN. The central body of the *Solar System*, a luminous sphere approximately 865,000 miles in diameter; it is believed to consist of a liquid inner portion with a gaseous outer covering. The average distance of the earth from the sun is about 93,000,000 miles. All forms of life on the earth and on other members of the solar system are dependent upon the radiation from the sun for their existence.

SUNRISE, SUNSET. The times at which the sun appears to rise above and set below the horizon, owing to the rotation of the earth on its axis. As the sun has a substantial apparent diameter, some time elapses between its first and last contacts with the horizon, and sunrise and sunset are therefore usually taken to be the times when its centre is on the horizon. Sunrise and sunset vary with latitude and with the *Declination of the Sun*.

SUNSHINE. The light received directly from the sun, which renders objects on the earth's surface visible. It consists of that part of the solar radiation with wave-lengths between those of red and violet light. In climatology and meteorology the duration of sunshine is used to indicate how sunny a place is; it is expressed in hours, often as the mean daily duration for a certain period. With sunshine are associated the ultra-violet rays, which are invisible and are of great therapeutic value. See *Isohel, Sunshine Recorder*.

SUNSHINE RECORDER. An instrument used for measuring the duration of sunshine each day. It usually consists of a spherical glass lens mounted at the centre of a framework in which a strip of graduated card is placed. The image of the sun produced by the spherical lens burns through the card, and the length of the burned track on the card, made as the sun travels across the sky, thus gives the numbers of hours of sunshine for the day.

SUNSPOT. A spot on the surface of the sun, sometimes so large that it may be observed, through a thin layer of cloud or dark glass, with the naked eye. Its origin is obscure, but it is believed to consist of a whirling mass of gas just within the sun's atmosphere. Some connexion between the number of sunspots and the occurrence of certain terrestrial phenomena has often been sought, but the only certain correlation is that between the number of sunspots and electrical and magnetic disturbances on the earth; when the former is a maximum, the displays of *Aurora* and *Magnetic Storms* also reach a maximum. There is a rough periodicity in the number of sunspots observed annually on the surface of the sun, the mean interval between maxima in the sunspot cycle being about eleven years.

SURF. Waves breaking into foamy water on the shore or around rocks.

SWALLOW-HOLE. See *Sink Hole*.

SWAMP. A tract of low-lying land which is saturated with moisture and usually overgrown with vegetation. It may adjoin a *Marsh* or a *Bog*, and there is sometimes confusion between the terms. A swamp (or marsh) is often formed when a lake basin fills up, for the surface is so flat that the *Run-off* of rain water is very slow, and in the damp soil vegetation grows and helps to maintain the swampy conditions. The term is sometimes applied to areas which are partly forest-covered, e.g. the Everglades of Florida and the Dismal Swamp of Virginia and North Carolina, U.S.A. By the drainage of swamps, e.g. in the Everglades of Florida, fertile agricultural land has been produced, for the soil has a high *Humus* content.

SWASH. The advance of sea water up a beach after the breaking of a wave. See *Backwash*.

SWELL. The regular, undulating movement of the surface of the sea, which does not break into waves.

Folding of Rock strata in a Syncline (trough) and an Anticline (arch).

SYNCLINE. The trough or inverted arch of a *Fold* in rock strata. See *Anticline*.

SYNCLINORIUM. A huge trough, in form resembling a *Syncline*, each limb of which consists of a number of small *Folds*.

SYNOPTIC CHART. See *Weather Chart*.

SYZYGY. The point at which two heavenly bodies are in *Conjunction* or *Opposition*; the term is usually applied to the moon, when it is in line with earth and sun, and thus coincides with new moon and full moon. See *Moon, Phases of*.

T

TABLELAND. A *Plateau* bounded by steep, cliff-like faces which lead abruptly down to the sea or the adjoining lowlands; South Africa provides the best-known example of a tableland.

TAÏGA. The *Coniferous Forest* land of Siberia, bordered on the north by the treeless, inhospitable *Tundra* and on the south by the *Steppes*. The principal species of trees are pine, fir, spruce, and larch. It contains many swampy areas, and during the spring much of the land is flooded by water from the upper reaches of northward-flowing rivers whose lower stretches are still frozen. The name is often applied, too, to other coniferous forests of the northern hemisphere, e.g. in North America.

TALUS* or SCREE. A mass of boulders and broken rocks of all sizes which accumulates at the foot of a cliff or mountain slope, having been broken from the main rocks by weathering and rolled down under the action of gravity. The talus slope is often so steep that a slight disturbance will send the whole mass sliding downwards.

TALUS CREEP. The gentle movement of rock fragments down a *Talus* slope.

TALWEG. See *Thalweg*.

TANKS. In most of the drier parts of peninsular India, the ponds or lakes formed by building mud walls across the valleys of small streams, so that water collects there during the wet season. The tanks may be used when the wet season is over, but during the hot season they dry up completely; in some years they do not fill even during the rains.

TARN. A small mountain lake, usually occupying a *Cirque*.

TECTONIC. Relating to the processes which tend to build up the various features of the earth's crust. The tectonic forces involved in moulding the earth's surface, for example, include those which break, bend, and warp the earth's crust, and create depressions and elevations; they are distinct from the forces of gradation, which tend to wear down the surface to a common level.

TEMPERATE ZONE. The middle latitudes, the zone between the *Torrid Zone* and the *Frigid Zone*: in the northern hemisphere between the *Tropic of Cancer* and the *Arctic Circle*, in the southern hemisphere between the *Tropic of Capricorn* and the *Antarctic Circle*. The sun is never overhead in this zone, but its rays vary considerably in directness, and there are therefore well-marked warm and cool seasons, or summer and winter. The term 'temperate' is appropriate to the middle latitudes of the southern hemisphere and to the oceans and the western coastal regions of the northern hemisphere but not to the interiors and eastern regions of North America and Asia.

TEMPERATURE. The degree of heat of a body, usually expressed in degrees on the *Centigrade Scale* or the *Fahrenheit Scale*, and measured by a *Thermometer*. On the earth, temperatures decrease in general from the equator towards the poles, though the temperatures of any particular place depend also upon other factors besides latitude – on altitude, proximity to the sea, prevailing winds, etc. Temperature also decreases with altitude. See *Lapse Rate, Range of Temperature, Isotherm.*

TEMPERATURE ANOMALY. The difference between the mean temperature of a place and the mean temperature along its parallel of latitude, both temperatures being reduced to sea level; the anomaly is positive if the place is warmer than the mean temperature along its parallel, and negative if it is cooler. The greatest temperature anomaly on the earth exists over the north-east Atlantic Ocean, including the British Isles, in January, when a considerable area has a positive temperature anomaly of over 10° C. See *Isanomalous Line.*

TEMPERATURE INVERSION. See *Inversion, Temperature.*

TEMPORALES. The strong SW. winds of *Monsoon* type blowing in summer on the Pacific coast of Central America.

TERRACE CULTIVATION.* An agricultural system by which mountain and hill slopes are cultivated: terraces are cut into the slopes, walls are built round the small level patches obtained, so that irrigation water and soil can be retained. Such terrace cultivation is common in mountainous areas where land is scarce or rainfall uncertain, e.g. in India, China, Java, Mediterranean countries, etc.

TERRACETTES or SHEEP-TRACKS. The small terraces often seen on steep grassy slopes, probably due to miniature landslips, and frequently used as paths by grazing animals.

TERRAL. The *Land Breeze* experienced on the west coasts of Peru and Chile. See *Virazon.*

TERRA ROSSA (Italian). A reddish clay soil found in limestone regions where the *Mediterranean Climate* prevails, e.g. on the bottom of a *Doline* or *Polje* or in a cave, being the insoluble residue left behind after solution of the limestone.

TERRA ROXA (Portuguese). A reddish-purple soil with a high *Humus* content which is found on the plateau of eastern Brazil; the land where it occurs has been extensively cleared for coffee plantations, e.g. in the state of São Paulo.

TERRESTRIAL MAGNETISM. See *Magnetism, Terrestrial.*

TERRIGENOUS. Derived from the land; for example, those marine deposits, derived from the land, which border the land masses, and include the sand, gravel, pebbles, etc., around the coasts, the sands, silts, and clays of the shallow waters, and the clays of the deeper waters on the fringe of the *Pelagic* zone, are known as terrigenous deposits.

TERRITORIAL WATERS. The belt of sea adjoining the coast, which is regarded as being under the territorial jurisdiction of the nation occupying the coast; a peaceful passage through the belt is normally permitted to the vessels of all nations. Most nations recognize the three miles' limit of territorial waters, but since World War II some countries, e.g. Iceland, have sought to extend this limit very considerably.

TETRAHEDRAL THEORY. A theory which seeks to explain the apparent symmetry in the distribution of the great land masses and the oceans, and suggests that as the earth cools its outer crust is warping towards a tetrahedral shape. In the western hemisphere, for instance, there is a large triangle of land, base to the north and apex to the south; in the eastern hemisphere, two triangular land masses unite at their bases to the north, and have separate apices in the south; there is a fourth land mass around the South Pole. These four triangular land masses represent the four points of the tetrahedron, projecting above the sea. Between them are three triangular masses of sea, bases in the south, where they unite, apices in the north.

THALWEG, TALWEG (German), or VALLEY LINE. The line joining the lowest points along a valley from the source of a river to its mouth: the concave curve thus produced marks the general slope of the river throughout its course.

THAW. The end of a *Frost*, when the temperature has risen above the Freezing Point, and ice and snow melt to water. In high latitudes the thaw is an annual event of great significance, taking place each spring, and leading to the melting of the snow and the liberation of ice-bound seas and rivers.

THEODOLITE. An instrument used for measuring angular distances in the vertical plane (elevation) and in the horizontal plane (azimuth); this is done by means of a telescope, whose motion can be accurately controlled. The instrument is used in surveying and also, for instance, in following a *Pilot Balloon*, when the direction and speed of the upper winds are determined.

THERMAL BELT. A zone in some of the valleys of the Appalachian Mountains, U.S.A., about 300 ft above the valley bottom, which is not affected by frost. The green colour of the vegetation is retained throughout the winter, and is separated by a distinct line from the frosted and blackened vegetation of the valley bottom, where temperatures have been much lower.

THERMAL EQUATOR. The imaginary line drawn round the globe through places which have the highest mean temperature for any particular period. Its position varies with the season, moving north in the northern summer and south in the southern summer, but it does not migrate so far north and south of the geographical equator as the sun itself. Being also influenced by the distribution of land and sea and the effects of ocean currents, its mean position does not coincide with the geographical equator, but lies to the north of it.

THERMAL SPRING. See *Hot Spring*.

THERMOGRAM. The continuous record of the air temperature, as measured by the *Thermograph*, usually to cover a period of a week.

THERMOGRAPH. A self-recording type of thermometer, in which a continuous trace of the air temperature is made on a *Thermogram* fixed to a rotating drum actuated by clockwork. One of the commonest forms uses a bi-metallic spiral, the expansion and contraction of which is transferred to a lever carrying a pen, and the latter then records the temperatures on the rotating thermogram. It is chiefly of value in recording variations in temperature rather than actual temperatures.

THERMOMETER. An instrument used for measuring temperature, usually by means of the change in volume of mercury or alcohol contained in a glass tube with a bulb at one end. In some English-speaking countries thermometers are constructed to the *Fahrenheit Scale*, elsewhere mainly to the *Centigrade Scale*.

THICKET. A small wood, or group of closely set trees, especially one with dense undergrowth.

THORN FOREST. A forest of small, thorny trees in a tropical or subtropical region, where the rainfall, about 10 to 20 inches annually, is too scanty to permit the growth of normal forests; during the long droughts the trees are leafless.

THRUST FAULT or REVERSED FAULT. A *Fault* in which the upper rock strata have been pushed forward over the lower strata, the upthrow being on the dip side of the *Fault Plane*. It may be formed by the breaking of an overturned fold, or *Overfold*, at the sharply bent crest of the *Anticline*.

THUNDERSTORM. A storm in which there are strong upward currents of air, forming well-developed *Cumulonimbus* clouds; from the latter violent showers of rain, possibly with hail, descend, while the development of static electricity causes flashes of lightning, and the thunder is produced by the expansion of air due to the tremendous heat of the lightning flashes. The first condition necessary for the development of a thunderstorm is a sufficient supply of moisture, in order that cumulonimbus clouds should form; the second condition is a high *Lapse Rate* of temperature, in order that there shall be a strong upward current of air for a depth of at least 10,000 ft above the cloud base. Thunderstorms are much more numerous and more severe in the equatorial regions than in the middle and higher latitudes, for in those regions there is, firstly, a greater supply of moisture; secondly, the more intense heat gives a high lapse rate, and causes the violent convection currents which, together with adequate moisture, produce the thundery cumulonimbus clouds.

TIDAL CURRENT. The movement of tidal water into a bay, estuary, harbour, etc., at the *Flood-Tide*, or out of it at the *Ebb-Tide*. Such currents are of great significance to a seaport like London, situated on a tidal estuary. In an estuary narrowing quickly inland, the tidal current may assume the proportions of a *Bore*.

TIDAL RANGE. The average difference in water level between high tide and low tide at a place.

TIDES. The alternate rise and fall of the surface of the sea, approximately twice a day, caused by the gravitational pull of the moon and to a lesser degree of the sun. In the open ocean, the amplitude of the tides, or the difference between the levels of high and low tides, is not more than one or two ft, but in the shallow seas bordering the continents it may be over 20 ft, and in narrow tidal estuaries it may be 40 or 50 ft. The amplitude at a place increases to a maximum, then decreases to a minimum, increases to a maximum again, and so on: see *Spring Tides, Neap Tides*. As the moon travels in its orbit in the same direction as the earth's rotation, a period of rather more than one day – 24 hours 50 minutes – elapses between successive occasions when the moon is vertically over a certain meridian; thus the interval between successive high tides or successive low tides is equal to half this period, i.e. about $12\frac{1}{2}$ hours. The tides may be regarded as two great waves, with their crests, 180° apart, lying along places where the moon is at the zenith or at the nadir, their troughs along places where the moon is on the horizon; the waves sweep round the earth as the latter rotates on its axis, and thus the crest, representing high tide, passes through any particular place twice every 24 hours 50 minutes. These movements are very much modified locally by irregularities in the coastline, e.g. by bays and gulfs.

TIERRA CALIENTE. In tropical Central and South America, the lowest of the three regions into which the land is sometimes divided according to altitude; it includes the hot, steamy, coastal plains and the slopes of the mountains to a height of about 3,000 ft. There is a luxuriant tropical vegetation, in the wetter areas dense forests; banana, sugar, and cacao plantations have been established, and on the mountain slopes maize, tobacco, and coffee are grown.

TIERRA FRIA. The highest and coolest of the three regions into which tropical Central and South America is sometimes divided according to altitude: the land between about 7,000 ft and 10,000 ft above sea level. It contains *Coniferous Forests*, which with decreased rainfall and lower temperatures gradually pass into *Scrub* and grassland.

The crops are those of the temperate lands, such as wheat, and there are considerable pastures. The mean annual temperature is about 10° C. below that of the *Tierra Caliente*; conditions are thus much more favourable to settlement, especially for Europeans, and the density of population in the tierra fria of Mexico is about four times that of the tierra caliente.

TIERRA TEMPLADA. In tropical Central and South America, the middle of the three zones into which the land is sometimes divided according to altitude: the area between about 3,000 ft and 7,000 ft above sea level. In general, temperatures are very uniform, and the annual range of temperature is often as low as 2° or 3° C. Maize, coffee, tobacco, etc. are cultivated, but owing to the light rainfall much of the area is valueless, even for pastures.

TILL. See *Boulder Clay*.

TIMBER LINE.* The boundary line above which trees do not grow. It is usually clearly marked on high ground in low and middle latitudes, but is not uniform in any region, for its height depends on local as well as general conditions of climate and soil. It is lower in the temperate zone than in the tropical zone, lower on the shady side than on the sunny side of mountains, and is highest on those slopes which provide the best protection from winds and the longest exposure to the sun.

TIME. See *Apparent Time, Local Time, Standard Time, Greenwich Mean Time*.

TOMBOLO. A *Bar* which joins an island to the mainland or joins two islands.

TOPOGRAPHIC MAP. A map on a sufficiently large scale to show the detailed surface features of an area, including its *Relief*, usually by means of *Contours*, and such physical features as forests, rivers, and lakes, and such artificial features as roads, railways and canals. Examples are the familiar Ordnance Survey maps of the United Kingdom on the scale 1 inch to 1 mile (1 : 63,360).

TOPOGRAPHY. A detailed description or representation of the features, both natural and artificial, of an area, such as are required for a *Topographic Map*.

TOPSOIL or TOP SOIL. Soil that is cultivated; the uppermost layer of true soil, as distinct from the *Subsoil*.

TOR. An isolated mass of rock, usually of granite, which has been subject to considerable weathering, and often assumes peculiar shapes; the granite rocks of Devon and Cornwall, England, provide good examples.

TORNADO. (1) In the Guinea lands of West Africa, a violent storm, often known as an 'African tornado'. It consists of a *Squall* blowing from a severe thunderstorm, often starting very suddenly and last-

ing only a short time, but it may cause considerable destruction on
land and sea; it is usually accompanied by torrential rain. It is
more liable to take place during the day, when the surface air is
strongly heated, and is most frequent during the transition periods
between the wet and dry seasons. The 'African tornado' is caused
by the meeting of warm, damp, monsoonal air from the south-west
with dry north-easterly or *Harmattan* air from the Sahara; it has
a *Front* several miles in length, sometimes extending to as much as
200 miles.

(2) An extremely violent *Whirlwind* covering a relatively small
area, its diameter being usually about one-quarter of a mile. It is
most frequent in the United States, east of the Rocky Mountains,
especially on the central plains of the Mississippi basin, where hot,
damp air from the Gulf of Mexico meets cool, dry air from the
north. It is a local phenomenon, travelling in an approximately
straight line at 20 to 40 miles per hour, and often dying out after
about 20 miles. While it lasts, however, it causes great destruction
along its narrow track; wind velocities near the centre are estimated to exceed 200 miles per hour, and trees are uprooted, buildings completely destroyed. The tornado is usually accompanied
by heavy rain and thunder. It occurs most frequently during spring
and early summer, almost always in the afternoon, when surface
heating reaches a maximum. In a suitable meteorological situation,
when an extensive area of low atmospheric pressure exists, several
tornadoes may develop on the same day.

TORRID ZONE. See *Tropics* (2).

TRADE ROUTE. A route, via land, sea, or air, by which trade is conducted. Overland the most important trade routes lie along the
railways, roads, rivers, and canals; over the oceans they are those
of the great cargo vessels, often following the *Great Circle Routes*; air
routes, for the carriage of light freight, mail, and passengers are the
most modern trade routes.

TRADE WINDS. The winds which blow from the sub-tropical belts of
high pressure towards the equatorial region of low pressure, from
the north-east in the northern hemisphere and from the south-east
in the southern hemisphere. In many areas they blow with extreme
regularity throughout the year, especially over the oceans, and
derive their name from the nautical expression 'to blow trade',
meaning 'to blow along a regular track'. In continental interiors
they blow much less steadily than over the oceans, but are fairly
regular over the hot deserts. Topography may cause local variations in direction, and even over the oceans there are variations;
in the North Atlantic, for instance, the direction changes from
north-east on the eastern margin to easterly on the western margin.

With the seasonal changes in the *Declination of the Sun*, the Trade Winds also move northwards and southwards, their range averaging about 5° latitude. Although weather in the Trade Wind regions is normally fine and quiet, *Tropical Cyclones* are often experienced there.

TRAMONTANA or **TRAMONTANO**. A cool, dry, northerly wind experienced in the Mediterranean region.

TRANSHUMANCE. The practice among pastoral farmers of moving their herds and flocks between two regions of different climate. In mountainous regions the animals are transferred from mountain pastures to valley pastures for the winter and back again to the mountain pastures for the summer; they are accompanied by herdsmen, sometimes by a considerable proportion of the local inhabitants, and a feature of the practice is that the latter occupy permanent dwellings in the mountains as well as in the valleys.

TRANSPORT. The carriage of goods from one region to another. In some ill-developed regions, e.g. China and Central Africa, it is still effected by human porters; various animals, including the horse, mule, donkey, ox, camel, elephant, reindeer, are also used. In more advanced countries, overland transport is chiefly by railway and road, with the steam locomotive and the motor vehicle respectively supplying the power, and rivers and canals are also utilized. Over the oceans the steamship and the motor vessel, and to a very small extent nowadays the sailing-ship, are employed, while air transport for light freight and mail is fast developing.

TRANSPORTATION. (1) The carriage of goods. See *Transport*.

(2) The process by which loose material from the earth's crust is conveyed from one region to another. Some of the agents which wear away the land also carry the material with them. See *Denudation*. Rivers are the chief agents of transportation, carrying material for immense distances, mud and sand being held in suspension and stones being rolled along the bed. Glaciers also bring material down from the mountains. See *Moraine*. The sea moves material from the coast into deeper water, and carries sand and pebbles laterally along the coast. The wind carries dust, and to a lesser degree sand, for great distances. Sometimes the transported material remains temporarily where it is deposited, either on land or in the sea.

TRANSVERSE VALLEY. A valley which, instead of being parallel to the mountain range, cuts across it. See *Longitudinal Valley*.

TRAVERTINE or **TUFA**. A deposit of calcium carbonate produced by a spring rich in lime; the calcium carbonate was originally dissolved by the water of the spring, with the help of carbon dioxide from the atmosphere. Travertine is commonly deposited from a *Hot*

Spring, e.g. at Mammoth Hot Springs, Yellowstone National Park, U.S.A. See *Sinter*.

TREATY PORT. In China, one of the seaports which were opened to foreign trade by treaty: foreign ships were permitted to load and unload there, and foreign merchants to live and own property. The most important of the treaty ports is Shanghai; Canton is another, and the remainder are situated mainly on the south-east coast.

TRELLISED DRAINAGE. The type of drainage produced by *Consequent Rivers* (parallel to the *Dip*) and *Subsequent Rivers* (parallel to the *Strike*), forming a rectangular or trellised pattern.

TRIBUTARY. A river or stream which contributes its water to a main river by discharging it into the latter, from either side, and at any point along its course. Each tributary does not necessarily increase the width of the main river, for often the additional supply of water is carried off by increased rapidity of flow.

TROGLODYTE. The inhabitant of a cave or rock shelter.

TROPICAL CYCLONE or TROPICAL REVOLVING STORM. A relatively small but very intense *Depression* which originates in tropical regions. Winds of hurricane strength circulate round the centre or 'eye' of the storm, their direction, as in the normal depression, following *Buys Ballot's Law*: in the northern hemisphere they circulate in an anti-clockwise, in the southern hemisphere in a clockwise direction. The 'eye' of the storm covers a restricted area, being often about 20 miles in diameter; here the atmospheric pressure is excessively low, and the air is sometimes calm and the sky clear. The whole cyclone has a diameter of 100 to 200 miles. The storm is accompanied by dense black clouds, torrential rain, and often thunder and lightning. The winds become less violent with increasing distance from the centre, but there is no abrupt transition to light winds. The cyclones follow fairly well-defined tracks, moving at an average speed of 10 to 15 miles per hour, travelling first westwards, then polewards, and finally turning eastwards. They originate near the western flanks of the oceans, where warm tropical currents supply an abundance of water vapour, especially in waters studded with islands; differential heating over land and sea probably causes a small area of low atmospheric pressure to develop. Most of them originate in the equatorial belt of calms, the *Doldrums*, when this belt is at its farthest limit from the equator; the necessary rotational movement can then form. They are thus most frequent in late summer and autumn. The necessary local conditions for development of a cyclone include a slack *Pressure Gradient* and very light winds, sultry weather, and an extensive area of falling atmospheric pressure. The tropical cyclone reaches its highest frequency in the China Seas, where it is known as a *Typhoon*. The other areas in which it is known are the Bay of Bengal, the southern Indian Ocean (Madagascar-Mauritius region), West Indies – where it is called a *Hurricane* – the SW. Pacific (Fiji-Queensland region), the Arabian Sea, and off NW. Australia – where it is called a *Willy-Willy*.

Losses at sea due to tropical cyclones are much smaller now than in the days of sailing ships. On land the storms tend to die out, but in passing from sea to land they often cause immense damage to coastal areas: buildings are destroyed; roads and railways are carried away; trees are uprooted and plantations ruined; moreover, thousands of lives have been lost, e.g. in China, Burma, and India.

TROPICAL GRASSLAND. See *Savanna*.

TROPICAL RAIN FOREST. See *Equatorial Forest*.

TROPICAL REVOLVING STORM. See *Tropical Cyclone*.

TROPIC OF CANCER. The parallel of latitude, roughly 23° N., indicating the extreme northern positions at which the sun appears directly overhead at noon. The sun's rays shine vertically on the Tropic of Cancer at the summer *Solstice* of the N. hemisphere.

TROPIC OF CAPRICORN. The parallel of latitude, roughly 23½° S., indicating the extreme southern positions at which the sun appears directly overhead at noon. The sun's rays shine vertically on the Tropic of Capricorn at the winter *Solstice* of the N. hemisphere, which is the summer solstice of the S. hemisphere.

TROPICS. (1) The *Tropic of Cancer* and *Tropic of Capricorn*.

(2) The region lying between the *Tropic of Cancer* and the *Tropic of Capricorn*, sometimes known as the Torrid Zone. Within the zone, the sun is directly overhead twice a year, its rays are never very oblique, and the weather in general, then, is always hot.

TROPOPAUSE. The boundary region in the atmosphere which separates the *Stratosphere* from the *Troposphere*.

TROPOPHYTE. A plant which acts as a *Hygrophyte* in one season, and a *Xerophyte* in the other. Trees in the *Savanna*, for instance, shed their leaves during the drought, temporarily becoming xerophytes, and produce them again with the onset of the rains, becoming hygrophytes.

TROPOSPHERE. The lower layers of the *Atmosphere*, i.e. those below the *Stratosphere*, extending from a height of about 6 miles to the earth's surface. Its depth, however, is very variable, being greatest at the equator, least at the poles. Within the troposphere the air temperature normally falls with increasing height; unlike the stratosphere, it is a turbulent, dusty zone, containing much water vapour and clouds.

TROUGH FAULT. See *Rift Valley*.

TROUGH OF LOW PRESSURE. An elongated area of low atmospheric pressure, normally forming an extension to a *Depression* or low pressure area, like a valley running upwards from a deep basin; a *Front* often lies along the middle of the trough. It is the opposite of the *Ridge of High Pressure*.

TRUCK FARMING. The term used in the United States to denote the *Intensive Cultivation* of vegetables for market, thus corresponding to the term *Market Gardening* which is usually employed in Great Britain. Truck farming, however, appears to be more specialized, and truck farms are often situated farther from the markets than are market gardens; the name may be due to the extensive use of trucks (lorries) for transporting the produce to market.

TRUE HORIZON. See *Horizon, Rational or True*.

TRUST TERRITORY. A territory administered for the United Nations by a member state after the end of World War II. Many Trust Territories are former *Mandated Territories*.

TSUNAMI. A large sea wave occasionally experienced along the coasts of Japan and in other regions, especially in the Pacific Ocean, caused by an *Earthquake* taking place on the ocean bed; it rises to greater and greater heights as it approaches the coastline, and on a single occasion has been known to cause the deaths by drowning of thousands of people. It is sometimes wrongly called a tidal wave.

TUFA. See *Travertine*.

TUFF. A rock consisting of *Volcanic Ash* which has been thrown out by a volcano during eruption. While tuffs of recent origin are usually loose, the older tuffs in general have been cemented together.

TUMULUS. See *Barrow*.

TUNDRA. The treeless plains of northern North America and northern Eurasia, lying principally along the Arctic Circle, and on the northern side of the *Coniferous Forests*. There is no corresponding region in the southern hemisphere. For most of the year the mean monthly temperature is below Freezing Point, and winters are long and severe, the ground being covered with snow. The summers are short and warm, but even in July the mean monthly temperature does not rise above 10° C.; relatively high temperatures may be reached during the daytime, but the subsoil, i.e. the ground about a foot below the surface, is perpetually frozen. This, and the strong, intensely cold winds of winter, make tree growth impossible. In summer, mosses and lichens appear in abundance and some flowering plants; much of the flat ground, where drainage is poor, then becomes swampy.

TWILIGHT. The faint light which illuminates the various regions of the earth before sunrise and after sunset; the interval during which the atmosphere is illuminated before sunrise (often known as the dawn) and after sunset. It is due primarily to the reflection of the sun's light from the upper layers of the atmosphere on the earth, while the sun itself is below the horizon. Morning twilight commences when the sun is 18° below the horizon, while evening twilight ends when the sun is also 18° below the horizon; each of these is termed the *astronomical twilight*. *Civil twilight* is recognized as beginning or ending when the sun is about 6° below the horizon, and is determined by the amount of light necessary to permit outdoor work. The duration of twilight is plainly much shorter in tropical regions than in higher latitudes. At midsummer there is a zone between the 48½° parallel and the Arctic Circle (or Antarctic Circle) where twilight extends throughout the night for a certain period. The number of nights of continuous twilight increases polewards.

TYPHOON. The *Tropical Cyclone* of the China Seas. It is experienced most frequently during the late summer and early autumn. The Philippine Islands lie in the direct track of many of the typhoons, and the coastal areas of southern China are also seriously affected. Like all tropical cyclones, the typhoons bring winds of tremendous strength and torrential rain, and destruction is often widespread, though the storms weaken as they cross the coastline and move inland. During the Swatow typhoon of August 1922, for instance, the sudden change of wind direction as the centre passed caused a huge wave to sweep over the town, and 50,000 lives were estimated to have been lost.

U

UBAC (French). A mountain slope which is shaded from the sun's rays owing to the fact that it faces more or less polewards; the contrast in light and warmth with a sunny slope is thus considerable. See *Adret*. The term is principally used in the Alps. The corresponding Italian term is opaco, the German schattenseite.

UMBRA. The central complete shadow of the earth or the moon in an *Eclipse*; the term is also applied to other phenomena, e.g. to the central dark region of a *Sunspot*.

UNDERTOW. The returning undercurrent from a wave breaking on the seashore.

UPLAND. The higher land of a region, in contrast with the valleys and plains.

U-SHAPED VALLEY. See *Valley*.

UVALA. In a *Karst Region*, a depression larger than a *Sink Hole* or a *Doline*, and often formed by the coalescence of several dolines, extending in diameter to a kilometre or more; it is, however, smaller than a *Polje*.

V

VADOSE WATER. Water that lies between the *Water Table* and the earth's surface.

VALE. A valley, usually of a broad, level type.

VALLEY. A long, narrow depression in the earth's surface, usually with a fairly regular downward slope. A river or stream usually flows through it, having carved it out from the surface rocks, and it is then known as a *river valley*: except in arid regions, the valley and the watercourse normally end together in the ocean, a lake, or another river. When a valley is young, it is narrow and its sides are steep; if the land is high, it has a steep gradient, unless it is very far from the sea, and it becomes deep. At this stage it is somewhat V-shaped in cross-section, and its tributaries are short. When a valley is mature, it is wider, its sides are often gentler, and its tributaries are longer. An old valley is very wide, has a broad *Flood-Plain* and a slight gradient; in fact, its surface is reduced almost to a plain, or to a *Peneplain*. A valley carved out by a glacier is U-shaped in cross-section. See *Longitudinal Valley*, *Transverse Valley*, *Rift Valley*, *Dry Valley*, *Hanging Valley*.

VALLEY LINE. See *Thalweg*.

VALLEY TRAIN. The deposit of rock material carried down by a stream originating from the melting ice of a glacier, and thus formed in a similar way to an *Outwash Plain*; unlike the latter, however, it is not spread over a wide area, but is confined within the walls of a valley.

VALLEY WIND. The flow of air up a valley during the day, often in still weather and with clear skies; the movement is due to the difference between the heating of the mountain slopes and the plains below, and frequently alternates with the nocturnal *Mountain Wind*. See *Anabatic Wind*.

VARIATION, MAGNETIC. See *Declination, Magnetic*.

VARVE. A pair of sediments, consisting of a lower coarse layer, usually silt, and an upper finer layer, usually clay, deposited in a lake, or the sea, from inflowing streams. By counting the varves, it has been possible to estimate the number of years since the end of the last *Ice Age*, for the coarser layer was the deposit of spring and summer, when the ice of the marginal lakes thawed, and the finer layer that of autumn and winter, when suspended material slowly sank, the lake surface above being frozen; each varve thus represents the deposit of exactly one year. The sediments are known as varved clays or varved sediments.

VEERING. The clockwise change of direction of a wind, e.g. from E. through SE. to S. It is the opposite change of direction to *Backing*.

VEGA. In Spain, irrigated land from which only one crop per year is obtained, in contrast to the *Huerta*.

VEIN. A crack or fissure in a rock which contains mineral matter deposited from solution. Metallic ores often occur in veins, and mines are located round them; much of the world's gold, silver, lead, etc., is found in such situations. The ores of these metals do not usually fill the cracks, but are often associated with a much larger amount of valueless mineral matter, which must be worked over to extract the ores which are valuable. The term is also applied to any layer of rock which has some economic value, e.g. a vein of ironstone. See *Lode*.

VELD (Afrikaans). Elevated, open country in South Africa, especially in the Transvaal, which might be considered suitable for grazing, forming part of the great *Tableland*. It is often divided into the High Veld, standing about 5,000 to 6,000 ft above sea level, the Middle Veld, and the Low Veld, standing about 1,000 to 3,000 ft above sea level. The High Veld consists of rolling, treeless *Grasslands*, broken here and there by *Kopjes*, and corresponds to the *Steppes*, *Prairies* and *Pampas*. The Middle Veld is marked by *Rands* as well as kopjes, and these are often covered with scrub. Rands and kopjes also occur in the Low Veld, though the region is generally uniform and is covered with scrub. Besides being divided according to altitude, as above, certain sections of the veld are sometimes classified according to soil, e.g. sand veld, or natural vegetation, e.g. bush veld.

VENDAVALES. Strong, squally, south-westerly winds experienced in the Straits of Gibraltar and off the east coast of Spain. Being associated with depressions, they blow chiefly in winter, and bring considerable rain.

VENTIFACT. A stone or pebble which has been shaped by wind-blown sand, usually in a desert, so that its surface consists of flat facets with sharp edges. The term dreikanter is sometimes used for ventifacts, but it wrongly suggests that the pebbles always have three facets.

VERANILLO. In tropical America (e.g. Colombia), a short, dry season which breaks the rainy season. See *Verano*.

VERANO. In tropical America (e.g. Colombia), the long dry season. See *Veranillo*.

VERNAL EQUINOX. See *Equinox*.

VIRAZON. The *Sea Breeze* experienced on the west coasts of Peru and Chile. See *Terral*.

VISIBLE HORIZON. See *Horizon, Sensible or Visible*.

VITICULTURE. The culture of the grape vine.

VOLCANIC ASH. Fine particles of *Lava* which have been ejected from a *Volcano* in eruption. The particles are rather coarser than those of *Volcanic Dust*, but the two terms are sometimes used interchangeably.

VOLCANIC BOMB. A lump of *Lava*, usually rounded in shape, which has been thrown out of a *Volcano* in the liquid state, solidifying as it fell. It varies in diameter from a few inches to several feet.

VOLCANIC CINDERS. See *Lapilli*.

VOLCANIC DUST. Fine particles of *Lava* which have been thrown out by a *Volcano* in eruption, having been blown into small particles by the force of the explosion. Sometimes the dust is shot high into the air, and is then carried immense distances by the wind.

VOLCANIC NECK.* A rocky crag consisting of solidified *Lava* which formerly filled the central opening of a *Volcano*, and has been left isolated when the remainder of the cone has been worn away by *Weathering*.

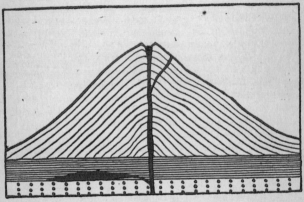

Section through a Volcano, showing volcanic neck, sill, and dyke.

VOLCANO.* A vent in the earth's crust caused by *Magma* forcing its way to the surface; molten rock, or *Lava*, is finally ejected, sometimes with explosive force, rock fragments and ashes being thrown into the air. The emissions of lava, or *eruptions*, often cause the volcano to take the form of a conical hill or mountain; the latter is gradually built up of ejected material, which is deposited most

thickly round the outlet. Eruptions take place from the top or sides
of the cone. They may be due to the production of steam, formed
when water percolating through the ground meets the hot magma:
dense clouds of steam generally accompany a volcanic eruption,
and most volcanoes are situated near to the sea. They may be due,
too, to the forces set up by earth movements. A volcano is regarded
as *active*, *dormant*, or *extinct*. When it is not in action, the pipe leading
down from its *Crater* is usually plugged with solidified lava. Of the
solid, liquid, and gaseous substances ejected from a volcano during
an eruption, the liquid materials are of greatest significance, forming the streams of lava which flow out of its crater or out of open-
ings in its side. With some volcanoes the overflow of lava takes place
quietly, and generally covers a wide area; Mauna Loa, on Hawaii,
is of this type. Others erupt with explosive violence; Mont Pelée, on
the island of Martinique, was of this kind: in 1902 it destroyed the
entire city of St Pierre and almost all the 28,000 inhabitants.
Others are of an intermediate type, liable at times to erupt explosively; Vesuvius, in Italy, provides an example. Volcanoes are
situated along the lines of weakness in the earth's crust, one of
which runs right round the Pacific Ocean; there is thus a chain of
active and recently extinct volcanoes along the eastern rim of
Australasia and Asia, continued down the western rim of North
and South America. See *Mud Volcano*.

V-SHAPED DEPRESSION. A term formerly applied to a *Trough of
Low Pressure*, in which the isobars take the shape of a letter V.

V-SHAPED VALLEY. See *Valley*.

VULCANICITY. The range of processes by which *Magma* rises into the
earth's crust or is ejected on to its surface.

W

WADI. A desert watercourse which is usually dry, and contains water
only occasionally, after heavy rainfall. The wadis generally lose
themselves in the desert; they may originate in the desert itself, or
in the melting snows of distant mountains. The term is mainly
used of the Sahara and the Arab countries of south-west Asia. See
Arroyo, *Nullah*.

WALLACE'S LINE. A line, first described by the British naturalist
A. R. Wallace, which divides the islands of Borneo and Bali on one
side from Celebes and Lombok (Indonesia) on the other side; the
native fauna on the two sides are so different that the line is held to
separate two distinct regions, the Oriental and the Australian, in
Zoogeography.

WARM FRONT. The boundary line at the earth's surface between a mass of advancing warm air and the cooler air over which it rises. The frontal surface between warm and cold air rises at a slighter angle than in the case of the *Cold Front*. The movement of the warm air over the cooler air leads to the formation of considerable cloud and usually precipitation ahead of the front, and after it has passed there is generally either slight rain or drizzle or no precipitation at all. The passage of a warm front through a place is thus normally marked by a rise in temperature, a slackening or cessation of precipitation, and a veer of wind. Warm fronts occur mainly in high latitudes, especially during winter, when the *Depressions* with which they are usually associated are most prevalent.

Upward movement of warm air at a Warm Front (broken arrow), and direction of movement of the front.

WARM SECTOR. The region of warm air lying between the *Cold Front* and the *Warm Front* of a typical *Depression*. As the depression moves along, the cold front gradually overtakes the warm front, amalgamating with it to form an *Occlusion*, and the warm sector thus often disappears at the earth's surface, though still existing in the upper air.

WARM WAVE. An abnormal rise in temperature experienced in temperate regions when warm air is introduced from lower latitudes, either ahead of a *Depression* moving eastwards towards the area or to the west of an *Anticyclone*. The term is principally used in the United States, where such warm waves occur during the summer in the central and eastern states.

WATERFALL.* A sudden fall of water, usually caused by a bed of hard rock in a river course, either horizontal or dipping gently upstream, overlying softer rocks. The softer rocks beneath are worn away by the river, so that the hard rock overhangs, and thus pro-

duces the waterfall. The layer of hard rock is undermined by the
falling water, and from time to time blocks of it are broken away;
the waterfall thus slowly recedes upstream, its crest always remain-
ing on the edge of the layer of hard rock. The best-known example
of this retreat of a waterfall is provided by Niagara Falls. A waterfall is also formed when a vertical mass of resistant rock extends
across the bed of the stream; in this case the fall does not recede,
but is gradually lowered as the surface of the hard rock is worn
down. A waterfall is formed, too, when a stream descends from a
Hanging Valley. Waterfalls impede navigation, but are of great
value in providing *Water Power*. See *Rapids*.

WATER GAP. A narrow gorge cut by a stream through a ridge of
hard rock.

WATER HEMISPHERE. That half of the globe which consists principally of ocean, lying mainly south of the equator; it is the opposite
hemisphere to the *Land Hemisphere*, has its centre near New Zealand
and contains only about one-seventh of the total land surface of the
globe.

WATER-PARTING. See *Watershed*.

WATER POWER. Energy obtained from natural or artificial waterfalls, and used either directly by turning a water wheel or a turbine,
or indirectly by the generation of electricity in dynamos driven by
turbines. See *Hydro-Electric Power*.

WATERSHED, WATER-PARTING, or DIVIDE. The elevated boundary
line separating the headstreams which are tributary to different
river systems or basins. It often has an extremely irregular course,
and does not necessarily follow the ridge of a range of hills or mountains. As rivers encroach upon each other's territory (see *River
Capture*), any watershed should be regarded as temporary. American
geographers use the term watershed as synonymous with *River
Basin*, employing water-parting or divide to denote the boundary
line between river systems. A *continental divide* is a line which separates the rivers flowing towards opposite sides of a continent.

WATERSPOUT. A *Tornado* occurring at sea, or the cloud associated
with it, more often observed in the tropics than in higher latitudes. A
portion of cloud shaped like an inverted cone appears to descend
from a heavy *Cumulonimbus* cloud till it meets a cone of spray raised
from the sea, and a column or spout of water is formed between sea
and cloud. The waterspout may be several hundred feet high, and
may last up to half an hour. Strong winds circulate round it in its
immediate neighbourhood. Its top often travels at a different speed
from its base, it becomes oblique or bent, and finally breaks and
disappears.

WATER TABLE or LEVEL OF SATURATION. The surface of the

Ground Water, or the surface below which the pores of a rock are saturated with water; this surface is uneven, and it is also variable. When rainfall is protracted, the pores of the rock become filled with water, some of which emerges at the junction of the permeable or porous rock and the underlying impermeable rock in the form of springs. As the water at the base of the porous rock runs out, the water table falls. During wet weather the water table rises, during dry weather it falls, but there are usually limits above which it never rises and below which it never falls. The lowest level to which the water table sinks in a certain locality is known as the *permanent water table*.

WEATHER. The condition of the atmosphere at a certain time or over a certain short period, as described by various meteorological phenomena, including atmospheric pressure, temperature, humidity, rainfall, cloudiness, and wind speed and direction. Near to the equator, the weather changes so little that it is almost synonymous with *Climate*.

WEATHER CHART or SYNOPTIC CHART. A chart or map on which are plotted details relating to the weather over the area represented; these details refer to a definite instant of time, e.g. 1200 hours G.M.T. (12 o'clock G.M.T.), and are collected at meteorological stations in the form of coded weather reports from a large number of observation posts. A continuous series of charts is plotted relating to certain recognized times during the day, usually at six-hourly or three-hourly intervals. When all the available observations for the area have been plotted, the isobars are drawn, usually at intervals of 2 millibars, and the main meteorological features inserted – *Depressions*, *Anticyclones*, *Fronts*, etc. Such a weather chart is often known as a synoptic chart, for it presents a synopsis of the meteorological situation at a certain time. Weather forecasts are prepared from the study of a sequence of such charts.

WEATHERING.* The disintegration of the earth's crust by exposure to the atmosphere; it is one of the main processes of *Denudation*, and may be classified as *mechanical* or *chemical weathering*. One of the most effective agents of disintegration is rain water containing carbon dioxide from the atmosphere. This solution, sometimes known as carbonic acid, readily dissolves the calcium carbonate of *Limestone*; this is an example of chemical weathering. Rain water containing oxygen from the air acts on ferriferous minerals. Another effect of rain in weathering may be observed in the formation of *Earth Pillars* (mechanical weathering). The wind also plays a part in weathering. Again, the heat of the sun causes rocks to expand, and the subsequent cooling causes them to contract, the alternate expansion and contraction leading to the cracking and breaking of the rocks near the surface. This effect is most noteworthy in the hot deserts, where the temperature changes are usually rapid. In higher latitudes frost is a more powerful agent. Rain water fills the cracks and pores of the rocks, and expands on freezing, exerting a great pressure on the rocks; the alternate thawing and freezing in time break up the rocks.

WEDGE OF HIGH PRESSURE. A *Ridge of High Pressure*, in which the isobars are V-shaped.

WELL. An underground source of water which has been rendered accessible by the drilling or digging of a hole from ground level to the *Water Table*; the water may then be raised or pumped to the surface. If a well is sunk to the water table of the wet season, but not as far as the lowest or permanent water table, it will contain water during the wet season but not during the dry season; if it reaches the permanent water table, however, it will always contain water. See *Artesian Well*. The term is also used in connexion with oil deposits, i.e. oil well.

WESTERLIES. The westerly winds which blow with great frequency in regions lying on the poleward sides of the sub-tropical high pressure areas or *Horse Latitudes*; the areas in which they blow are often known as the 'regions of the Westerlies'. In winter they move southwards in the northern hemisphere, affecting the Mediterranean regions and much of the horse latitudes, bringing winter rain to those areas; in summer they move northwards again. From about 40° N. to the Arctic Circle, and from about 35° S. to the Antarctic Circle, however, they blow throughout the year. The winds derive their name from the prevailing direction; in the northern hemisphere this is mainly south-westerly, in the southern hemisphere north-westerly. Unlike the *Trade Winds*, they are very variable in force and direction, especially in the northern hemisphere, where the winds may be easterly for a considerable period. The weather in their area is marked by an almost constant procession of *Depressions* and *Anticyclones* moving eastwards, and is made more complex in the northern hemisphere by the alternation of oceans and continents; on the eastern side of Asia, for instance, the effect of the westerlies is broken by the *Monsoons*. In winter the North Atlantic Ocean is probably the stormiest region of the globe, at any season, but in summer the winds and weather are much less violent. In the southern hemisphere, on the other hand, the westerlies blow with great strength and regularity throughout the year over the almost unbroken expanse of ocean, and have given the name *Roaring Forties* to the region.

WET BULB TEMPERATURE. The temperature registered by a wet bulb thermometer, an ordinary thermometer whose bulb is enclosed by a small piece of muslin which is kept constantly moist by a wick dipping into a small water-bottle. The thermometer thus registers the atmospheric temperature reduced by evaporation from a moist surface, the degree of cooling depending upon the rate of evaporation. The difference between the readings of the dry bulb thermometer and the wet bulb thermometer in relation to the air temperature gives a measure of the *Relative Humidity*, and tables have been prepared for interpreting the readings; a very slight difference indicates the extreme dampness or high relative humidity of the air, for there is then very little evaporation from the surface of the wet bulb thermometer, and a considerable difference indicates the comparative dryness or low relative humidity of the air. See *Hygrometer*.

WET SPELL. In the British Isles, a period of at least fifteen consecutive days each of which has had .04 in. of rain or more; the definition has not been internationally accepted. See *Dry Spell*.

WHEAT BELT. That part of the *Prairies* of North America on which wheat is extensively cultivated; it lies east of the Rocky Mountains, and stretches from northern Alberta in Canada to northern Texas in the U.S.A., and is the most important wheat-growing area in the world. In recent years the United States has been second and Canada third in world production (the U.S.S.R. being first).

WHEAT CRESCENT. The crescent-shaped area extending round the drier, western margin of the South American *Pampas*, on which wheat is widely cultivated; it has enabled Argentina to become one of the leading world exporters of wheat.

WHIRLPOOL. A circular eddy in the sea or in a river produced by the configuration of the channel, or by the meeting of two currents, or from similar causes.

WHIRLWIND. A small column of air rotating rapidly round a vertical or slightly inclined axis, having an area of low atmospheric pressure at its centre. It sometimes extends to a considerable height, and in a desert may cause a *Duststorm*, or, on a smaller scale, a *Dust Devil*.

'WHITE HORSES' or WHITECAPS. The broken white crests of sea waves, produced by the wind.

WHITE MAN'S GRAVE. The name formerly applied to the lands along the Guinea coast, West Africa, more especially to the district around Freetown, capital of Sierra Leone. The atmosphere is almost perpetually hot and humid, the winds being mainly from the Gulf of Guinea; the drier and healthier *Harmattan* from the north-east rarely reaches this coastal area. In view of the progress made in recent years in combating tropical diseases, the name is no longer applicable.

WILLIWAW. A violent *Squall* experienced in the Strait of Magellan, where the winds are almost constantly strong and westerly.

WILLY-WILLY. The *Tropical Cyclone* experienced over north-west Australia, mainly in late summer. It originates in the hot Timor Sea, and moves first south-westwards, often causing severe losses to the pearl fishing industry; then it recurves to the south-east, crosses the coast, where it causes more damage and brings heavy rain, and moves overland towards the Great Australian Bight. In the interior the willy-willy soon loses its violence, and it is generally welcomed there for the rain that it brings. The term is also applied to *Whirlwinds* that occur over the Australian deserts – including the relatively small *Dust Devils*.

WIND. A current of air, moving with any speed in any direction. The direction of a wind is indicated by the point of the compass from which it blows: e.g. a south wind is one which blows from the south. The speed of a wind is usually given in English-speaking countries in miles per hour, at sea in *Knots*. Winds may be classified according to speed, e.g. on the *Beaufort Scale*, and the actual velocities are measured by the *Anemometer*.

WIND GAP or **AIR GAP.** A narrow gorge, originally a *Water Gap*, from which the water that formerly flowed through has disappeared.

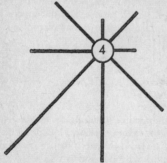

Simple type of Wind Rose.

WIND ROSE. A diagram illustrating the proportion of winds which blow from each of the main points of the compass at a certain place, taken over a considerable period; it thus shows the *Prevailing Wind* at a glance. In a simple form of wind rose, the proportion of winds blowing from the eight principal points of the compass is indicated by straight lines of varying length and true direction converging to a small circle; a number in the circle gives the percentage of cases in which the air was calm. In some forms of wind rose, the different forces of wind are also indicated, e.g. by varying the thickness of the lines.

WIND VANE. An instrument used to indicate the direction of the wind. It usually consists of a freely moving horizontal arm with a pointer at one end, and below this a fixed framework indicating the four *Cardinal Points* of the compass.

WINDWARD. The side or direction which faces the wind.

WOOD. An extensive group of trees, or a small *Forest*.

X

XEROPHYTE.* A plant which is adapted to living in a region where little moisture is available, i.e. where *Drought* conditions normally prevail. Its structure is modified in various ways in order that the maximum amount of water shall be obtained and conserved: roots are long or enlarged, leaves are small and thick, or even absent altogether, as in many cactus plants, being sometimes replaced by thorns; the stems are sometimes fleshy, enabling the plants to store water for extremely long periods; a thick bark or a coating of wax may protect the plant against transpiration.

Y

YARDANGS. In the deserts of Central Asia, the steep-sided, undercut, overhanging ridges of rock, separated by long corridors which have been cut out by wind-borne sand; they are roughly parallel, and lie in the direction of the dominant wind.

YAZOO RIVER. A *Tributary* which is prevented from joining the main river because the latter has built up high natural *Levees*; it therefore runs parallel to the main river for some considerable distance, eventually joining it much farther downstream. The name is derived from the Yazoo River, tributary of the Mississippi, the classic example of this type of stream.

YEAR. The period of time taken for the earth to complete one revolution in its orbit round the sun, or the period taken to complete the cycle of the *Seasons*. Measured from one vernal *Equinox* to the next, it equals 365 days, 5 hours, 48 minutes, 46 seconds; for ordinary purposes it is taken to be 365 days in length, and a *Leap Year* of 366 days is introduced every fourth year. See *Calendar*.

Z

ZENITH. That point in the heavens, i.e. the *Celestial Sphere*, which is vertically above the observer. The *astronomical zenith* is defined as the point at which a plumb-line, extended upwards from the earth's surface, would intersect the celestial sphere; the *geographical zenith* is defined as the point at which the line perpendicular to the earth's surface intersects the celestial sphere. The two points do not always coincide.

ZENITHAL or **AZIMUTHAL PROJECTION.** The type of *Map Projection* in which a portion of the globe is projected upon a plane tangent to it; it may thus be regarded as a special case of the *Conical Projection*, in which the cone is so flattened that it finally becomes a plane. As the name 'azimuthal' indicates, on this projection all points have their true compass directions from the centre of the map. The tangent plane on which the projection is made is not always tangent at the pole; in fact, it is usually tangent at some other point, which is to be the centre of the map. See *Zenithal Equal-Area, Zenithal Equidistant, Stereographic, Gnomonic, Orthographic Projections*.

ZENITHAL EQUAL-AREA PROJECTION. A *Zenithal Projection* in which any area on the map truly represents the corresponding area on the globe. See *Homolographic Projection*.

ZENITHAL EQUIDISTANT PROJECTION. A *Zenithal Projection* in which distances from the centre of the map truly represent the corresponding distances on the globe.

ZODIAC. That zone of the heavens in which lie the paths of the sun, the moon, and the chief planets. It is divided into 12 signs (the Signs of the Zodiac), each sign being 30 degrees in extent counting from the position of the sun at the spring *Equinox*, and is marked by 12 constellations. Each of the 12 months of the year is thus associated with one of the zodiacal constellations.

ZONDA. In Argentina and Uruguay, a hot, sultry wind bringing tropical air from the north, analogous to the Australian *Brickfielder*, and usually preceding the *Pampero*. It has a very enervating effect on the inhabitants.

ZONES. The five regions or belts into which the earth is sometimes divided. They are the areas within the *Arctic Circle* and the *Antarctic Circle*, called respectively the North and South *Frigid Zones*; the area between the Arctic Circle and the *Tropic of Cancer*, and the area between the Antarctic Circle and the *Tropic of Capricorn*, known respectively as the North and South *Temperate Zones*; and the area on both sides of the equator between the Tropics of Cancer and Capricorn, known as the *Torrid Zone*. This division is geographically inexact, for it takes no account of variations due to altitude, proximity to the oceans, etc., and a much more useful division is that into *Climatic Regions* or *Natural Regions*. The term is also widely used in the sense of 'regions' of various types, e.g. *Abyssal* zone.

ZONE TIME. See *Standard Time*.

ZOOGEOGRAPHY. The study of the distribution of animals on the earth's surface.

ZOOPHYTE. An animal which resembles a plant, e.g. a *Coral Polyp*, a sponge.

*Some more books published by Penguins
are described on the
following pages*

MINERALS IN INDUSTRY

W. R. Jones

This book, now fully revised and brought up to date, with many important additional sections, provides in very readable form information about almost all the minerals and mineral products used in industry.

The introduction gives a brief and clear account of the chief ways in which the minerals of commerce are formed. Then follow, in alphabetical order, descriptions of all these (except of mineral substances used as fuels) with an account of their uses in industries and the arts. Particulars are given of the chief mineral-producing countries and of their importance in relation to world production.

Diagrams specially drawn for this book show the world average production for most of the minerals during a selected five-year normal period, and outline world maps at the end show graphically the location of the chief sources of many of the minerals which are as essential to industry as food is to man.

THE ANCIENT EXPLORERS

M. Cary and E. H. Warmington

Two eminent scholars – both professors in the University of London – collaborated in the writing of this volume, which remains a standard work on ancient travel and discovery. Concerned with the actual journeys made rather than with the geographical speculations of ancient scholars, they tell how, before Arabic expansion closed the gates of the Mediterranean Sea, men had coasted Western Europe and penetrated the continent south of the Danube and Rhine, sailed from Suez to Canton and probed deeply into Asia, and – even if they failed to circumnavigate Africa – had been as far as Sierra Leone and Port Delgado. They describe, too, the objects of these journeys, the crude equipment of sailors, and the scanty geographical knowledge on which they proceeded.

In this 'Hakluyt' of the ancient world one reads – often with surprise – of Greeks in India and Romans in China, of the account of the source of the Nile given by one Diogenes, and of Pytheas's extended visit to the boorish inhabitants of Britain. And it is clear that the Great Age of Discovery, in the fifteenth and sixteenth centuries, was heavily indebted to these more ancient explorers.

A HISTORY OF LATIN AMERICA

George Pendle

About Latin America and its problems there prevails what *The Economist* recently called an 'awe-inspiring ignorance'. An authoritative and concise introduction to an area of such great economic potentiality is certainly needed.

This history has been written by a specialist who has been closely connected with Latin America for the last thirty years. In tracing the development of civilization from the earliest times down to Fidel Castro – with the principal countries taking their place in the continental story – the author helps to place current events in their context.

Many races and classes have contributed to the civilization of this great serpentine land-mass, with its vast mountain ranges, rivers, prairies, forests, and deserts: Indians, European *conquistadores*, priests, planters, African slaves, *caudillos*, liberal intellectuals, commercial pioneers.

Simón Bolívar once classed the Latin Americans as a distinct variety of the human species, and in discussing their characteristics George Pendle identifies two notable qualities: the 'El Dorado spirit' and an inclination to prefer people to 'things and rules'.

A DICTIONARY OF MODERN
HISTORY
1789–1945

A. W. Palmer

This book is intended as a reference-companion to the personalities, events, and ideas of the last century and a half. While the prime emphasis is on British affairs and on political topics, the Dictionary is intended to represent trends in the history of all the major regions of the world. Particular care has been taken to include numerous entries on the U.S.A. and on Russia, areas which earlier books tended to neglect. The entries are arranged in alphabetical order (with appropriate cross-references) and are in essay form, ranging in length from little more than 100 words to nearly 2,000. There are entries on economic, social, religious, and scientific developments, but not on the Arts. Explanations are given of some of the famous descriptive phrases of the period. About a third of the entries are biographical.

The book is intended as an aid to study, and not a substitute for it. The author hopes that it will explain the passing allusion and stimulate an interest in unfamiliar facets of historical knowledge.

UNDERSTANDING WEATHER

O. G. Sutton

This book now revised and brought up to date, with a new chapter on long-range forecasts, reflects the change in the approach to meteorology. Once largely descriptive, it is now rapidly becoming an exact science.

It outlines in simple terms the main features of climate and weather. The mysteries of warm and cold fronts, of isobars, anticyclones, and the rest, are all exposed. In addition, the book explains the *methods* which are needed to analyse the particularly complex data of meteorology, and shows how electronic computers will help in the future.

Can man forecast the weather a long way ahead, or even perhaps control it? Do nuclear tests affect it? These and many other relevant questions are discussed. The author is a professional meteorologist who seems to justify his claim that this subject is 'the most fascinating of all the earth sciences.'

GEOLOGY AND SCENERY IN ENGLAND AND WALES

A. E. Trueman

Scenery depends on land structure – in other words, its geology. Everyone interested in the countryside, how it has taken shape, why it presents us with the varied beauties of mountain and moorland, river valley and fertile meadow, is, if often unconsciously, appreciating its geology.

Sir Arthur Trueman's book – first issued in 1938 as *The Scenery of England and Wales* and later reprinted in revised form as a Pelican book – makes it abundantly clear that geology is pre-eminently a layman's science. The author believes that the geologist acquires an eye for country and an understanding of nature not excelled by that of the artist or the poet.

The English and Welsh countryside is considered district by district, each chapter dealing with one type of country – the Cotswold Stone belt, the Chalk lands, the Pennine Moors, the Lake District, and so on. The reader already familiar with the areas dealt with will learn to view them in a new light, while those who seek information about the peculiar features and delights of localities yet unknown to them will be amply rewarded. The author began to write it while himself on holiday, and has kept the needs of other holiday-makers prominently in mind.

There are 95 illustrations and diagrams, specially drawn to illustrate the text.

APPLIED GEOGRAPHY

L. Dudley Stamp

Geography, literally 'writing about the earth', still means to far too many of us, influenced by school-day memories, the wearisome descriptions of countries in which lists of capes, bays, mountains, rivers, towns, and products play a major part. But its real interest is to describe and reflect the physical build and the natural resources, the sequence of human occupation and social organization, which have built the world we know, and will change and develop it in years to come. To know and understand these causes and their certain or probable effects is vital in all planning for the future; and this is the field of applied geography. In this book, a pioneer effort in its field, the principles of geographical survey and analysis are applied to the problems of Britain today.

THE FACE OF THE SUN

H. W. Newton

A special job of astronomers during the International Geophysical Year, when this book first appeared, was the study of the surface of the sun. The tools which modern technical skill has newly placed at their service, among them radio telescopes such as that at Jodrell Bank, are adding almost daily to our knowledge of the constitution and the activity of the one star we can observe at what, in terms of stellar distances, may be called 'close range'.

Mr Newton tells the story of what men have learnt about the sun during the three-and-a-half centuries since Galileo's early telescope first disclosed the existence of sunspots. The fifty-thousand-mile-long tongues of gas which shoot out from the sun's surface into space, the streams of solar particles which enter earth's atmosphere to cause magnetic storms that disturb our radios, and to paint the night sky with displays of the Aurora Borealis, the obscure effects of solar activity on our weather conditions are described, discussed, and so far as is yet possible explained.

And – because our sun, we may suppose, is in many ways typical of many of the countless stars of remotest space – we are given a glimpse of happenings far beyond our own system, in similar stars at the remotest bounds of the universe.

THE FACE OF THE EARTH

G. Dury

The young natural science of geomorphology – the study of the form of the ground – is much less forbidding than its name. It is developing fast, and already promises to achieve some independence both of geology and of physical geography. In this book a professional geomorphologist tells how knowledge in this field is advancing, examines some of the hotly-disputed problems which have to be solved, and discusses the processes by which construction and erosion affect the physical landscape. Among the topics receiving attention are the weakening of rocks by weathering, their removal by the forces of erosion, the cyclic development of the land-surface, the evolution of river-systems, the effects of volcanic action and of glaciers, and the surface forms of deserts.

In choosing his examples, the author has been able to select freely from the results of his own field work. There are 102 diagrams in the text and 48 pages of plates.

PENGUIN REFERENCE BOOKS